D0088862

THE VISIBLE OPS HANDBOOK

STARTING ITIL IN 4 PRACTICAL STEPS

Information Technology Process Institute

KEVIN BEHR, GENE KIM AND GEORGE SPAFFORD

The Visible Ops Handbook: Starting ITIL In 4 Practical Steps

Copyright 2004, IT Process Institute (ITPI)

First Edition published June 2004.

"Visible Ops", "ICOPL" and "IMCA" are trademarks of ITPI. ITIL is a registered trademark of the UK Office of Government Commerce and not the ITPI. Capability Maturity Model, CMM, CERT, and CERT Coordination Center are registered in the U.S. Patent and Trademark Office by Carnegie Mellon University. All other trademarks and company names are the property of their respective owners.

For more information, please contact:

ITPI
2896 Crescent Avenue
Eugene, Oregon 97408
Main Telephone: (541) 485-4051
Main Fax: (541) 485-8163
http://www.itpi.org
info@itpi.org
ISBN 0-9755686-0-4

Credits

Copy Editors: Crystal Behr and Ron Neumann
Graphics and Production Manager: Harold Metzger
Project Coordinator: Mary Matthews
Technical Editor: Tom Good
Book Design: Integrity Design and Marketing

Acknowledgements

The ITPI offers a special thank you to a number of important contributors. Without the knowledge, hard work, and dedication of the following individuals, we would be challenged to produce the important tools which help to shape IT for our time.

Scott Alldridge
Julia Allen
Ruby Christina Bauske
Brandon Casey
Grant Castner
Steve Darby
Jeremy Epstein
Ruby Gates
William Hertling

Charles Hornat
Joe Judge
Rich Llewellyn
Dwayne Melançon
Craig Morgan
Bill Murray
Stephen Northcutt
Nevin Oliphant
Fred Palmer

Mike Prospect
Jackie Shaffer
Bill Shinn
Troy Thompson
Jan Vromant
Dan Waite
Henry Wojcik
Ron Zika

Testimonials:

"A frequent complaint of ITIL consultants is that not many ITIL implementation tools are publicly available. For the experienced IT Service Management practitioner, it sometimes seems we have to re-invent the wheel each time. Visible Ops fills a big part of that void. It provides a practical insight in how to kick-start an IT Service Management improvement effort. Its common sense approach and very readable style give this book a mandatory place in the library of any IT manager. Visible Ops is indeed comparable to the manual for an Emergency Room of a hospital. I particularly liked the fact it does not pretend to be an operational 'bible' for the phases beyond the ER.

Visible Ops describes four steps to control an IT environment. The unassailable logic behind these steps is based on the practical experience of the authors, Gene Kim and Kevin Behr. These same steps can easily be mapped to any maturity model and Visible Ops hence describes a roadmap to maturity.

The first two phases of Visible Ops help organizations control the infrastructure. The third Visible Ops phase helps organizations control the services, in the spirit of Service-Oriented Architectures and IT Service Management. The last phase of Visible Ops helps organizations control the strategic value, which provides an opportunity for IT to align itself with the business and to gradually maximize its 'bang for the buck.' The easy mapping between the Visible Ops phases and any maturity model validates the compelling logic of the book."—JAN VROMANT, ITSM CONSULTANT

"Gene and Kevin have hit the proverbial 'IT nail' right on the head. When I educate customers on the benefits of documented and repeatable procedures such as ITIL and COBIT, they are always concerned about the complexity and where to start. Visible Ops creates a logical starting point and details the key 'issues and indicators.' This handbook is a 'must read' for IT Managers and Directors who are implementing a mandate from their CIO or Board of Directors to become compliant for auditors and federal regulations." –HENRY E. WOJCIK, DIRECTOR, ENTERPRISE SERVICE MANAGEMENT, NETWORK DATA SYSTEMS, INC

"The Visible Ops Handbook is the Rosetta Stone that the IT industry and its leadership have been seeking to allow them to communicate the value of ITIL to the business. Visible Ops is simple and clear, provides a roadmap of how to make an IT department not only perform better, but also to deliver more value back to the business. Without doubt, each of the four steps they outlined have value and are well supported."—DANIEL S. WAITE, SENIOR CONSULTANT, BMC SOFTWARE

"The Visible Ops handbook provides a great roadmap for IT executives to see their way through the thicket of chaotic operations and into the clearing of repeatable processes. It follows in the footsteps of software development processes like the Capability Maturity Model (CMM), and offers the potential to provide a real ROI by reducing the effort in wasted firefighting."—JEREMY EPSTEIN, SENIOR DIRECTOR, PRODUCT SECURITY, WEBMETHODS, INC.

"Finally, a 'best practice' that is based upon research and industry knowledge. Too often, best practice papers are written with little or no influence or research from those actually performing the work in the real world. This approach is a step-by-step, methodical approach for any organization looking to get a grip on Change Management and improve operations. The books format and introduction of methodologies will set the pace for all future publications!"—CHARLES HORNAT, GLOBAL INFORMATION SECURITY MANAGER

"Visible Ops provides the IT practitioner at any level with a catalytic approach to improving operational controls. The Visible Ops toolset helps organizations find a toehold in spite of sheer cliffs of chaos. If you are looking to start or improve configuration management, champion a repeatable server provisioning process, and institute meaningful metrics that breed quality decisions, Visible Ops is the place to start. I recommend this to any IS Management, as well as any senior management with a technical background or IT staffers with management ambitions."—BILL SHINN, SECURITY ENGINEER

"If you are in IT, you are most likely currently dealing with issues of firefighting due to rampant changes, or are deliberately ignoring them due to lack of time and resources. Visible Ops provides a clear-cut methodology and steps to effectively deal with these issues. This book provides a scaleable template that fits around any size shop to get back in control, and then actually stay there. They show you how to regain control of critical changes, whether it's an entire data center rebuild, a single device failure, or upgrading an entire server farm to a new software release, and then continue to manage them effectively from deployment to production to retirement." —TROY THOMPSON, ITIL CERTIFIED CONSULTANT

"This is a very valuable resource for anyone just getting started. If this resource had been available when I was putting together the Change Management plan for our department, it would have saved me many hours of research. I highly recommend it as both a reference and developmental tool. It will help you identify the processes and order in which you should develop and implement the various ITIL BS 15000 process areas. More importantly the tips for audit preparation will help you identify the specific areas of improvement and help you identify and target areas requiring an organizational culture change. Well written, easy to follow, with good examples; It has everything you need from beginning development through the measuring the results."—JACKIE SHAFFER, SYSTEMS PROJECT ADMINISTRATOR, FLORIDA DEPARTMENT OF EDUCATION

"In general, this book provides a synopsis of the techniques and methodologies we at SIAC use to provide close to 'five nine' uptime for our owners and customers."—MIKE PROSPECT, VICE PRESIDENT, SECURITIES INDUSTRY AUTOMATION CORPORATION

"Visible Ops is a methodology that comprehensively responds to major issues I have raised over and over again in my long career in financial and technology auditing. To attest to the reliability of systems, auditors need to see: controls in place, controls documented, controls communicated, and evidence of the controls in action. Visible Ops shows IT managers how to build their operational processes so they can answer the auditors' eternal question: 'How do we really know?'"—RUBY CHRISTINA BAUSKE, LEAD TECHNOLOGY AUDITOR, CPA, CIA, CISA, CISSP

"Change management done wrong is painful, cumbersome and results in needless firefighting. However, effective change management done correctly enables IT operations and information security to work more efficiently and better support the business objectives. Furthermore, it makes audits easier to pass and perform. The Visible Ops book clearly explains in a practical and manageable approach what it takes for organizations to implement change management that really works. If organizations agree to follow the approach in this book and stick to it, they will see how structured and disciplined change management will actually make their lives easier, will not stifle responsiveness or flexibility, and will help to extinguish many of the fires."—CRAIG MORGAN, CISSP, PRINCIPAL SECURITY CONSULTANT, ENSPHERICS (A DIVISION OF CIBER)

"As an IT consultant, I continually deal with many people from different disciplines who are smarter than me. Even so, they often ask me which books I rely on to do my job. The Visible Ops book just became one of them—in my toolkit, I'd always want a pocket knife, a can of Sterno, a compass and this guide."—RON ZIKA, ITSM CONSULTANT

Foreword

I remember well my first substantive conversation with Gene Kim in March 2003. We were in Orlando, Florida, sitting at the bar after a full day of mind-numbing security conference presentations. During this exchange, I found out what Gene was up to, and I had one of those proverbial light bulb moments. What if we could find a way to define mature IT operational processes and then embed well-defined security controls within these processes? If we could do this, we could make great progress in addressing security in the normal course of operational business, instead of by individual heroics. What made this promising and exciting was that Gene had seen this in action, and had studied how certain organizations that he called "best in class" were not only doing it, but doing it exceptionally well.

Gene introduced me to his partner in crime, Kevin Behr. We found that we had similar interests, and embarked on finding a useful way to work together. By July of 2003, we had a collaboration agreement in place between Carnegie Mellon University's Software Engineering Institute and the IT Process Institute. In October 2003, we co-hosted the first Best in Class Security and Operations Roundtable at the SEI, bringing together leaders and high performers in IT operations and security.

On this journey, I have learned the following from Gene and Kevin:

1. They have a unique ability to observe, analyze, and synthesize information and experiences from organizations operating across a wide range of market sectors. In doing this repeatedly, they have created value, resulting in strong, long-term relationships of trust with their clients and partners.

2. They have identified critical characteristics of what it means to be high performing in IT operations and security, as evidenced by an organization's culture, beliefs, behaviors, capabilities, and actions. They have observed how high performing organizations view the problems as well as the solutions. This handbook codifies much of this work.

3. They passionately believe in, and have begun to demonstrate, the power of mature process definitions to bring about stability and control in complex IT environments, including the requirement for auditable and verifiable controls.

4. They have invested significant time and energy in their own education (and mine) and in building a rich value network of leading and respected professionals in IT operations, security, and audit to assist and advise this work.

5. Their observations and experiences (and those of their clients and partners) on the current state of IT operations and security are remarkably similar to those of the software development community before the existence of a body of community-accepted software development process definitions (as captured in the SEI's Capability Maturity Model® for Software).

Why has the SEI embarked on this journey with ITPI? We share a mutual desire to improve the condition of IT operations and security. These capabilities do not stand alone; they live in an enterprise context. The tougher aspects of improvement are in people and process, even though the community at large still tends to view localized technology solutions as the path for improvement. We share the belief that sustained improvement requires the creation of an executive-level community of practice, who will integrate the goals and objectives of IT operations, security, audit, risk management, process management, project management, and governance. All of these capabilities are required to bring about an operational environment that can deliver repeatable, predictable, defined, secure, measurable, and measured operational processes, thereby achieving operational excellence.

We share the objective of helping organizations make common sense common practice. By addressing the difficult questions, "How and where do you start?" this handbook is a significant step in the right direction.

Julia Allen
Senior Member of the Technical Staff
Carnegie Mellon University, Software Engineering Institute
Networked Systems Survivability Program, home of the CERT® Coordination Center

Table Of Contents

Introduction

Practitioners in information technology (IT) face pressures on many fronts. In addition to the demands to become more efficient, IT must now address challenges to maintain a secure state and comply with regulatory requirements. For example, the Sarbanes-Oxley Act of 2002 is forcing publicly held U.S. corporations to attest to the fact that internal controls are both in place and effective. IT operational best practices, such as the Information Technology Infrastructure Library (ITIL), provide a framework to start defining repeatable and verifiable IT processes. However, as organizations attempt to use ITIL to begin their journey towards process improvement, they face two very difficult questions: How and where do you start?

This handbook provides an overview of the methodology that we have developed known as "Visible Ops." Since 2000, we have met with hundreds of IT organizations and identified eight high-performing IT groups with the highest service levels, best security, and best efficiencies. What was most amazing about them was that they shared the following attributes: a culture of change management, a culture of causality, and a culture that fundamentally valued effective and auditable controls, promoting fact-based management. Visible Ops reflects the lessons learned about how these organizations work and describes a control-based entry point into the world of ITIL that others can leverage to springboard their own process improvement efforts.

In the IT industry, Stephen Elliot, an IT Senior Analyst with IDC, showed that on average, 80% of IT system outages are caused by operator and application errors.[1] This motivated our need to dig into causal factors of infrastructure downtime, which continually revealed shortfalls in change management practices. Often, many organizations would have well-documented change management practices, but in reality, no one ever followed them. In many of these cases, the goals and motivations for having change management were not clear to management or to the practitioners themselves. Another key finding was that having a documented change management process was necessary, but far from sufficient, to achieve high-performing characteristics. In the high-performing organizations we studied, change management was embedded in their culture, and had a very different meaning than in typical organizations. This book is dedicated to describing those practices that set the high-performers apart.

Something Must Need Improvement—Otherwise, Why Read This?

> *"The most likely way the world will be destroyed, most experts agree, is by accident. That's where we come in; we're computer professionals. We cause accidents."*—NATHANIEL BORENSTEIN

The motivation for ITIL, change management, and overall process improvement is well known. The trade press is full of stories about cost cutting measures, outsourcing, and regulatory requirements from Sarbanes-Oxley, HIPAA (The Health Insurance Portability and Accountability Act of 1996), BASEL II, FISMA and so forth. The list of people talking about the problems is already large enough, so we promise to keep the discussion of the problem domain to a minimum. In this booklet, the issues and challenges that we address include:

---

[1] Source: Stephen Elliot, Senior Analyst Network and Service Management, IDC, 2004. Note, additional information can also be found from the Gartner Group at http://www4.gartner.com/DisplayDocument?id=334197&ref=g_search

- Organizations have change management processes, but view these processes as overly bureaucratic and diminishing of productivity. There must be more to change management than bureaucracy, good intentions and scarcely attended meetings.

- Organizations where, deep down, everyone knows that people circumvent proper processes because crippling outages, finger-pointing, and phantom changes run rampant.

- A "cowboy culture" where seemingly "nimble" behavior has promoted destructive side effects. The sense of agility is all too often a delusion.

- A "pager culture" where IT operations believes that true control simply is not possible, and that they are doomed to an endless cycle of break/fix triggered by a pager message at late hours of the night.

- An environment where IT operations and security are constantly in a reactive mode, with little ability to figure out how to free themselves from fire-fighting long enough to invest in any proactive work.

- Organizations where both internal and external auditors are on a crusade to find out whether proper controls exist and to push madly for implementing new ones where they are not in place.

- Organizations where IT understands the need for controls, but does not know which controls are needed first.

What You Do And Do Not Need To Know

You do not need an extensive knowledge of ITIL, process improvement, security or audit to benefit from this book. These topics are introduced in this handbook as they become necessary in the Visible Ops methodology. Our intent is to create a working knowledge of critical concepts in these domains, both to serve as a primer and to introduce the language necessary to work with other functional departments, such as security and audit. However, we recognize that each one of these domains is an entire vocation and field of expertise unto itself, so we list recommended resources in the appendices for those wishing to learn more. An evolving list of resources can be found on the ITPI Web site at http://www.itpi.org.

Structure Of The Book

This booklet presents information in the following order:

- Visible Ops: What is it and why does it work?
- There are four Visible Ops phases. In each, we describe:
  - Issues and indicators
  - Specific prescriptive steps to solve the issues
  - Benefits and what you are likely to hear as the steps are implemented
- Appendices that provide a brief primer on auditable controls, information on how to proactively prepare for an audit, a summary of ITIL, and other helpful resources.
- In each of the Visible Ops phases, "Helpful Tips When Preparing for an Audit" sections appear in grey call-out boxes, highlighting areas of special interest to those who interact with auditors.

Please note that we use "production" and "operations" interchangeably to specify the team primarily responsible for day-to-day infrastructure operations and maintenance. Specifically, this does not include the release management team. They are in the preproduction portion of the service delivery lifecycle.

Visible Ops

> *"It is not enough to show that a situation is bad; it is also necessary to be reasonably certain that the problem has been properly described, fairly certain that the proposed remedy will improve it, and virtually certain that it will not make it worse."*—ROBERT CONQUEST

We developed the Visible Ops methodology because everyone seemed to be asking the same urgent question: "I believe in the need for IT process improvement, but where do I start?" There were no satisfactory answers. Although ITIL provides a wealth of best practices, it lacks prescriptive guidance: In what order and how should the practices be implemented? Moreover, the ITIL books remain relatively expensive to distribute widely. The third-party information that is publicly available on ITIL still tends to be too general and vague to effectively aid organizations. This booklet provides step-by-step guidance and a prescriptive roadmap for organizations starting or continuing their IT process improvement journey. Visible Ops uses ITIL terminology, and is intended to be an "on-ramp" to the rest of the ITIL body of knowledge.

History Of Visible Ops

Since early 2000, Gene Kim, CTO of Tripwire, Inc., and Kevin Behr, CTO of IP Services, have studied what contributes to the success of high-performing IT organizations. IP Services is a business process outsourcing company, managing thousands of servers for Fortune 50 organizations. At IP Services, the IT operations group reports to Kevin, and for years, he tried to understand how to best increase service levels and decrease cost to maximize value. Tripwire is a software vendor for a product that detects change—it was originally written by Gene in 1992 as an intrusion detection technology to help system administrators recover from the 1988 Morris Internet Worm. Gene has spent years trying to understand why their largest customers kept insisting that Tripwire's software was not a security technology, but instead, a technology to enforce their change management processes.

Kevin and Gene began working together when they discovered they had a common passion to really understand what differentiated high-performing IT organizations from their more typical counterparts. Visible Ops began to take shape when they started studying a list of organizations that Gene had been keeping for years, which he called "Gene's list of people with amazing kung fu."

After years of research and investigation, Kevin and Gene now refer to this list more formally as "the high performing IT operations and security organizations with the highest service levels, as measured by mean time to repair (MTTR), mean time between failures (MTBF), and availability;[2] the early integration of security requirements into the operations lifecycle; the lowest amount of unplanned work; and the highest server to system-administrator ratios." What makes the organizations on this list especially astonishing is that they also have more efficient cost structures than lower-performing organizations.

To coordinate and expand their efforts, their works were donated to the Information Technology Process Institute (ITPI). The ITPI is a not-for-profit organization engaged in three principle areas of activity: research, benchmarking and the development of prescriptive guidance for practitioners and business executives. The ITPI has collaboration agreements in place with research organizations such as The University of Oregon Decision Sciences program and The Software Engineering Institute at Carnegie Mellon University. The ITPI also attracts many other contributors through the ITPI Community of Practice List (ICOPL). At the time of writing, there are

---

[2] Appendix D is the glossary of terms.

hundreds of top practitioners from IT Operations, Security, Audit, Management, and Governance on the ICOPL, representing thousands of years of IT experience.

Through research, development and benchmarking, the ITPI creates powerful measurement tools, prescriptive adoption methods (such as Visible Ops), and control metrics to facilitate management by fact. The end result of these efforts is to assist organizations with their IT process improvement efforts. This booklet serves as an example.

Common Characteristics Of High-Performing IT Organizations

What makes high-performing organizations so different from average organizations, both qualitatively and quantitatively? We observe that high-performing IT organizations share the following characteristics:

- **Server to system administrator ratios greater than 100:1**—This means that each system administrator controls more than 100 servers. In contrast, organizations not using effective processes see ratios around 15:1.

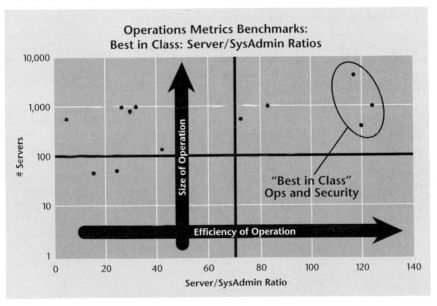

Figure 1: Server to System Administrator Ratio

- **Low ratio of unplanned to planned work**—Only 5% of operational expense goes toward unplanned work. From our ongoing benchmarking, we find that average organizations spend 25–45% of their total operational expenses on unplanned, unscheduled work.

- **Higher staffing early in the IT lifecycle**—Continual deployment of resources and staff in the preproduction build phase, where the cost of defect repair is least expensive.

- **Collaborative working relationships between functions**—IT operations and security work together to solve common objectives, with IT operations performing most of the work and security acting as coach and consultant.

- **Posture of compliance**—Trusted working relationship between IT operations and auditors, because controls are visible, verifiable and regularly reported on.

- **Culture of change management**—Ubiquitous understanding throughout the organization that changes must be managed in order to achieve business objectives.

- **Culture of causality**—Through the use of controls and metrics, these groups identify and solve problems through logical use of cause and effect, instead of a culture of "let's see if this works."

- **Management by fact**—These organizations value controls and metrics, not only to aid effective problem solving, but to aid fact-driven decision making, as opposed to "management by belief" or "management by the honor system."

Why Did We Use ITIL?

To understand what the best in class organizations were doing, Gene and Kevin wanted to determine the union and intersection of their IT processes. In other words, what are the common practices of all the high-performing IT operations organizations studied, and which ones are necessary to achieve the high-performing characteristics? Even this line of questioning was a challenge, because each organization had independently developed their own processes, and each had Darwinistically evolved to learn from past mistakes to prevent certain IT disasters from ever happening again.[3] Because they were building their own playbook, as opposed to using an external standard, each organization called similar processes by different names. For example, one organization's "change management" process was another's "work authorization request system" or "change control" process. As a result, Kevin and Gene first needed some way to normalize terminology in order to determine what processes these organizations had in common.

To resolve this terminology problem, they did a Google search on "release management and change management," which brought them to ITIL. ITIL is a compilation of IT best practices, provided without prioritization or any prescriptive structure. ITIL provides a framework and catalog of IT operational processes, distilled from thousands of man-years of experience. Initially created in the late 1980s, the ITIL body of knowledge continues to be enhanced and better organized, most significantly (in our opinion) in the form of the BS 15000, which divides all the ITIL disciplines into five key areas: Release Processes, Control Processes, Resolution Processes, Relationship Processes, and Service Delivery Processes.

---

[3] Similarly, FAA insiders say that "behind every regulation is an airline crash."

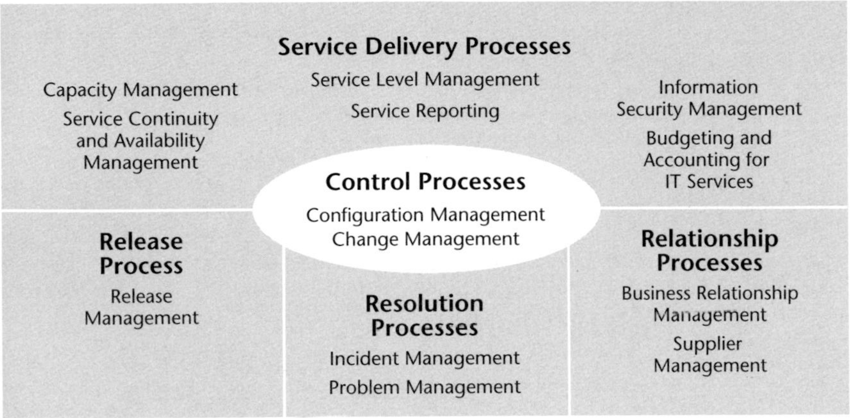

Service Delivery Processes

Capacity Management

Service Continuity and Availability Management

Service Level Management

Service Reporting

Information Security Management

Budgeting and Accounting for IT Services

Control Processes

Configuration Management
Change Management

Release Process

Release Management

Relationship Processes

Business Relationship Management

Supplier Management

Resolution Processes

Incident Management

Problem Management

Figure 2: BS 15000 view of ITIL process areas[4]

The BS 15000 categorizes the ITIL capabilities into five areas. Each are briefly described below:

- **Release Process**—This process area answers the question of "where does infrastructure come from before it is deployed?" This includes activities such as the planning, designing, building, and configuring of hardware and software. Unfortunately, release processes are traditionally the last process area that organizations invest in. Yet this is the process area that delivers the highest return on investment, because it encompasses the entire pre-production infrastructure, where the cost of defect repair is lowest.

- **Control Processes**—This process area covers maintaining production infrastructure, not only to prevent service interruptions, but also to efficiently deliver IT service. This is done through change management, as well as asset and configuration management. BS 15000 defines change management as well as asset and configuration management as primary controls. As Stephen Katz, former CISO of Citibank, once said, "Controls don't slow the business down; like brakes on a car, controls allow you to go faster."

- **Resolution Processes**—This process area is triggered when production infrastructure does go down, service is interrupted, or there is a security issue. Incident management owns the customer relationship, and problem management owns the tasks of turning each problem into a known error that can be more efficiently resolved the next time it happens. All too often, organizations that spend too much time firefighting are unable to spend time in the previous two process areas.

- **Relationship Processes**—This area focuses on the processes necessary to support effective customer relations as well as the management of third party vendors from a performance and contractual standpoint.

- **Service Delivery Processes**—The goal of these processes is to provide the best possible service levels to meet the business needs of the organization. This process area includes the monitoring and management of IT infrastructure as it relates to Security Management, Availability and Contingency Management, Capacity Management, Financial Management and Service Level Management and Reporting.

[4] BS 15000-1:2002—"IT Service Management: Part 1: Specification of Service Management." British Standards Institute. September 2002. Page 2.

In the high-performing organizations, the common processes were in the release, controls and resolution areas. All of the high-performers had repeatable and verifiable processes to provision infrastructure in a known good state. They had a culture of change management as a primary way to do work and they all used causality in their problem resolution processes. It is interesting to note that none of the high-performing shops were using ITIL at the time of the research. But again, ITIL provided a framework to name and normalize the practices that the high-performing organizations had in common.

ITIL is still not in a form where you can simply distribute the ITIL volumes to your entire IT organization and expect everyone to know what issues to tackle first and what everyone's role should be. Yet experienced IT practitioners who have built their own playbook of lessons and have learned from their own disasters, or near-disasters, are likely to love reading the ITIL volumes. They will see reflections of their own belief systems and management practices in the ITIL, and recognize the wealth of hard-won lessons and processes contributed by other IT practitioners that they can add to their playbook. With these expectations, ITIL can be a tremendous wealth of useful information.[5]

One last note on ITIL: We are continually awed and amazed that so many organizations have re-created the hard-won lessons embodied in ITIL over and over again. Because each of these organizations created their own methodology, when these IT operations organizations meet, even though they are doing very similar things, they cannot speak a common language. One of the first things that a community of practice must develop to share best practices is a common vocabulary. By using ITIL, we normalized the various terms into a standard framework.

In our opinion, just mapping your IT operational processes to ITIL has value. It allows organizations to share best practices plus leverage the tremendous wealth of ITIL and its various advocacy groups, such as the itSMF.[6] At an even more practical level, being aware of ITIL terminology facilitates interaction with other IT organizations and lowers the risk of misunderstandings.

Why Visible Ops Works

Since 2002, we have presented our research and the Visible Ops methodology to a wide cross-section of the IT community. Through this process, we have received positive feedback from hundreds of people in virtually every industry, company size and functional role. Sometimes we ask ourselves why Visible Ops resonates so well. We now believe that it is because Visible Ops is both logical and intuitive, equally accessible to technical and non-technical stakeholders. Typical reactions are: "This makes so much sense—I have to show this to my boss" and "Wow, our company needs to do this" and even "Visible Ops shows that common sense is rarely common practice."

By replicating how the high-performing IT organizations work, Visible Ops presents practices that not only make sense, but also can be implemented in any organization (i.e. "this really isn't rocket science"). For novice organizations, Visible Ops provides useful guidance on where to start their improvement efforts. For more mature organizations, Visible Ops provides a framework for continual improvement.

[5] For more information on ITIL, please see Appendix B.

[6] http://www.itsmf.com/

Visible Ops is also accessible to business management, security, and audit because it is controls-based. By being based on controls, not only are regulatory issues addressed, but controls help provide the reliable delivery of IT service. Visible Ops identifies key issues that undermine service levels and security, and provides prescriptive guidance to address them. These issues are:

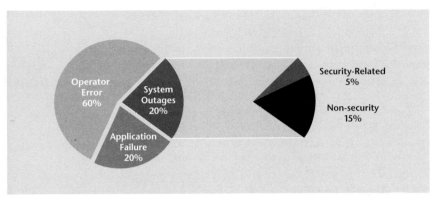

Figure 3: 2004 IDC Study on Causes of Network Downtime[7]

- **Human factors effect successful change**—Implementing a change management process and having it actually followed are two very different things. To meet the requirements of the business, effective change management is a necessity. In order for it to work, human factors must be addressed.

- **80% of outages are self-inflicted**—Donna Scott, VP & Research Director, Gartner, notes that, "80 percent of unplanned downtime is caused by people and process issues, including poor change management practices, while the remainder is caused by technology failures and disasters."[8]

- **80% of MTTR is often wasted on non-productive activities**—Determining the cause of an outage consumes a great deal of valuable time without effective change management. This protracts the outage and makes repair more difficult.

- **Absence of a "culture of causality"**—People often manage and work by intuition and "gut feel." Consequently, they fail to use problem-solving skills and causality to resolve issues. The Microsoft Operations Framework (MOF) study showed that their high-performing customers reboot servers 20 times less often than average and have five times fewer "blue screens of death."

- **Rebuild vs. Repair**—High-performing organizations make it easier to rebuild infrastructure than to repair it. The results are higher and more predictable service levels, plus, by rebuilding from documented standard builds, more junior staff can handle repairs.

---

[7] Source: Stephen Elliot, Senior Analyst Network and Service Management, IDC, 2004.

[8] Miller, David. "Hardware High-Availability Programs in Action. (Product Information)." ENT News. June 1999. http://www.entmag.com/archives/article.asp?EditorialsID=6753

Visible Ops—Key Characteristics

Visible Ops is neither a death march nor a monumental multi-year undertaking. In fact, we have seen organizations successfully complete the first three phases of Visible Ops in 90 days. The initial part of the methodology is broken down into manageable sub-projects prior to moving into a continuous improvement process. The goal is to create the fewest processes necessary to enable sustaining improvement. To do this, each of these sub-projects has the following characteristics:

• **Definitive Projects**—Each phase is a project with a clearly defined objective.

• **Ordered**—Each phase is specifically designed to build upon the previous phase.

• **Catalytic**—Each phase returns more resources to the organization than it consumed, thus fueling the next phase.

• **Auditable**—Each phase creates auditable processes that generate on-going documentation in order to prove controls are working and effective.

• **Sustaining**—Each phase creates enough value to the organization that the processes developed remain in place, even if the initial driving forces behind its implementation disappear.

This approach has many benefits. First, because of the relatively short length of each phase, concepts and their benefits are proven faster. Second, getting executive sponsorship and funding for four smaller phases is easier than for a big vision with a distant promised payoff.

Facilitating A Productive Working Relationship Between IT Operations, Security And Audit

All too often, IT operations groups have an unproductive working relationship with security and audit. Visible Ops creates a framework that creates productive interfaces between these groups, through repeatable, verifiable and auditable IT processes. By exposing IT controls and acceptance points, security and audit are able to review changes before they are implemented, and detect when these controls are circumvented. These controls are used not just to avoid circumstances which can lead to security incidents or unplanned work, but they also allow the continual monitoring and reduction of variance.

Bill Shinn, a System Security Engineer with a Fortune 100 financial institution, has studied the correlation between the amount of unplanned work and the number of security incidents. He has observed that as the number of unplanned changes increases, the likelihood of insecure configurations increases correspondingly, as do the number of incidents where security must investigate issues. For example, security may be called upon during a network outage because the issue is obviously "another firewall problem" instead of an undocumented change made by the network administrators. In contrast, when changes are planned, security has a chance to review, approve and respond to the changes early in the production lifecycle and can route issues to the responsible parties. This early involvement increases the overall IT organization's ability to fix systemic issues that lead to unnecessary firefighting and security problems.

Similarly, IT auditors often are exasperated by the absence of documented processes, the lack of a defined desired state, and an inability to attest to whether or not the current state meets the documented control objectives. Without these, auditors are unable to determine if risks and controls are in balance. In the absence of verifiable controls, they must go into "archaeology" mode and make a judgment on whether a material risk exists or not. This is not to say that the IT operations group is necessarily doing a poor job. Indeed, if the staff turnover is sufficiently low, the "tribal knowledge," or combined team knowledge, can compensate for a lack of formally documented processes, observes Ron Zika, a Senior Consultant for Waypoint, Inc. Visible Ops creates the instrumentation where auditors can review the processes and controls for effectiveness without having to enter into a forensics analysis mode. This leads to a more productive working relationship, smoother audits and less time spent on audit preparation and remediation.

Although IT operations, security and audit have very different roles, the three groups are often needlessly at odds because of the lack of effective controls. By improving processes and controls, all parties benefit by creating a more productive working relationship and allowing the groups to more efficiently achieve common business objectives. How this is done will be covered in more depth later in the book.

An Overview Of The Four Visible Ops Phases

Visible Ops gives organizations a means to begin their process improvement journey. After studying the high-performing organizations, we focused the methodology on four key phases:

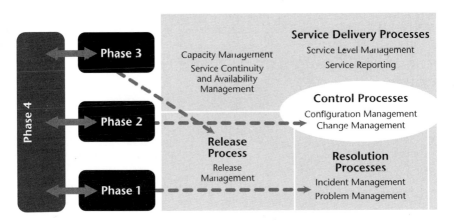

Figure 4: Visible Ops' Four Phases and Relevant ITIL Process Areas

- **Phase 1: "Stabilize the Patient"**—In this phase, we curb the number of outages by freezing change outside of scheduled maintenance windows. We also modify the first response process of problem managers by ensuring that they have all change related information at hand about what could have caused the outage.

- **Phase 2: "Catch & Release" and "Find Fragile Artifacts"**—Often, infrastructure exists that cannot be repeatedly replicated. In this step, we inventory assets, configurations and services to identify those with the lowest change success rates, highest MTTR, and highest business downtime costs. Fragile artifacts are identified and then treated with extra caution to avert risky changes and massive episodes of unplanned work.

- **Phase 3: Establish Repeatable Build Library**—The highest return on investment comes from implementing effective release management processes. This step creates repeatable builds for the most critical assets and services to make it "cheaper to rebuild than to repair." We take the priceless paintings identified in the previous step and work to create equally functional prints that can be mass-produced.

• **Phase 4: Enable Continuous Improvement**—The previous steps have progressively built a closed loop between the release, control and resolution process domains. This step implements metrics to enable the continuous improvement of all of these process areas to best meet business objectives.

Now that we've provided a brief overview, let's dive into the details of Visible Ops.

Visible Ops In Detail

Visible Ops focuses primarily on the effective management of change to begin process improvement efforts. Organizations have two means to embark on the journey. One method is to use the ITPI's Integrity Management Capabilities Assessment (IMCA) to identify weak areas and facilitate implementation planning (see Appendix C). The alternative is to simply follow the Visible Ops steps, fully completing each phase before proceeding to the next.

Note that Visible Ops applies to all infrastructure systems being managed by IT operations spanning servers, databases, routers, switches, firewalls, networking devices, storage systems and so on. Keep in mind that change affects all types of infrastructure—not just servers! At times (such as in the third phase) Visible Ops may seem focused on servers, but the management principles are broadly applicable to all types of IT infrastructure.

Phase One: "Stabilize The Patient"
And "Modify First Response"

Our goal in this phase is to reduce the amount of unplanned work as a percentage of total work done down to 25% or less. Organizations that are in a constant firefighting mode can have this percentage at 65% or even higher. The first phase of Visible Ops resembles the triage system used by hospitals to allocate scarce medical resources. In a similar fashion, IT must identify the most critical systems generating the most unplanned work and take appropriate action to gain control. The primary goal of this phase is to stabilize the environment, allowing work to shift from perpetual firefighting to more proactive work that addresses the root causes of problems.

Issues And Indicators

The issues and symptoms that we tackle in phase one are:

Issue	Narrative Example
Formal service levels and/or informal expectations are not being met.	"Despite having an availability target for 99% last quarter, we did not achieve it. Truthfully, we didn't even come close. Because of an especially horrendous outage that spanned almost two days on December 18, which generated all-nighters for everybody and lots of expensive overtime, our availability figures for the quarter came in at 94%."
IT is creating the majority of their own work through self-inflicted problems related to uncontrolled change.	"Obviously, 94% availability is not acceptable. We looked seriously at what caused this particular outage, so we could prevent it from happening again. Looking back at those horrible 48 hours, we now know that it was because one of our developers decided to upgrade 50% of the Web servers with some new code, changing 93 critical executables. It caused a certain Web shopping session to lock up the servers. The upgraded servers locked up so hard that they didn't even reboot—they were frozen with stack traces everywhere. Very bad news."
When systems are down, 80% of the MTTR is dominated by simply trying to characterize the outage and determine causal factors. Only 20% of the recovery time is spent actually repairing the infrastructure.	"When we were first hit by these failures, our entire site would go down, taking down our whole line of business. We convened an emergency meeting with everyone in Ops and R&D, and one of the first questions we asked was, 'Did anyone change anything?' Of course, the answer was, 'no.' Everyone swore that they didn't change anything. "So, while the business was losing approximately $20K each minute because we were in our peak holiday retail season, we went from pointing fingers at each other to eventually screaming at each other. Why the tension? Because this had happened before. No one takes accountability for changes, and without proof, all we had to go on was suspicion. This path was definitely not productive. Quite frankly, because these disasters had happened before, we all had a sense of déjà vu and lots of blame was being cast around."

Issue	Narrative Example
When changes are detected, who made the change and why are not always readily apparent. Sometimes changes have a very long fuse, detonating long after the change was made.	"Cooler heads eventually prevailed, and we started to piece together some useful clues. We discovered that there was a development upgrade being tentatively explored. It was nothing official, but it was a promising lead. We tracked down the development manager leading the effort, who told us that the responsible developer had just gone on vacation the previous night.

"This was terrible news. We had no idea whether he made any changes, and even if he did, what exactly did he change? How could we best unwind the changes to get these boxes back to a running state?

"Eventually, the development manager caught the developer at some airport via cell phone, and the developer admitted that he had done a 'small upgrade,' but was adamant that his change could not have caused the outage. He said that it was 'inconceivable' that his change would cause the failure—he actually used the word 'inconceivable!' Just like in the movie *The Princess Bride*!

"'Inconceivable.' Yeah, right. When we copied these supposedly harmless executables onto a test box, we were immediately able to replicate the problem. It may seem bad that we chewed up over 24 hours to get to this point, but then the real bad news hit us..." |
| System failures happen during very inconvenient periods, causing stress and damaging IT's reputation. | "Did I tell you that this outage happened four shopping days before Christmas? And that our company does 80% of our business in December? Did I tell you that we were unable to process orders on what should have been the busiest revenue day for our online Web commerce systems? Our executives were so furious that they actually brought in external auditors to figure out whose butt to kick. That pretty much catches us up to the current state of affairs.

"Auditors. Great. Well, I've had my butt chewed out before, and I've survived all the audits so far. But I'm pretty sure the executives brought in some consultants as well, to 'explore options.' At this point, everyone fears the worst: Outsourcing." |
| Typically, we have found that average organizations spend 35–45% of their time on unplanned and unscheduled work. | "We've had over ten almost-catastrophic failures this year. There may be good reasons that there is a feeling in the IT operations staff that things are in utter chaos. For whatever reason, this is definitely not the information that is being presented to management. In fact, I've seen some of the IT reports being presented to the executives, and it actually shows us doing a good job: 99.9% availability, no major failures, etc."

"The crazy part is how much they're focusing on the wrong numbers to come up with reports that say what a great job we're doing! When it comes down to it, we are spending too much time on crisis management and falling behind on projects the rest of the time. When SQL Slammer hit us one weekend, organization-wide, we probably spent over $35 million on unplanned work. Because it was a weekend, it didn't hit the availability numbers, but how can we ignore all that unplanned work?" |
| It is all too easy for one change to undo a previous change or even a whole series of changes. | "So you'd think that after the SQL Slammer disaster, we would have learned our lesson, right? I wish that were true. Unfortunately, after all the emergency patching, in the next quarterly build, half of the servers that we deployed didn't have one of the critical patches installed, and the same thing happened to us less than two months after the initial SQL Slammer attack. We looked like complete goofballs, and frankly, I'm wondering why someone hasn't been canned for this!" |
| Unnecessary problem management escalation due to invisible failed changes. | "Not only are we spending more time on unplanned work than planned work, the majority of the unplanned work is self-generated. Everyone is so bad at making configuration changes of any kind that I look forward to any time when we are not allowed to make changes. In fact, our developers create so much carnage during our application code upgrades that here in IT operations, we sometimes joke about crashing the developers' systems so we can get some real work done.

"When our critical app servers go down, the first thing we say to ourselves is, 'Hey, I bet a developer just did a code push!'" |

Issue	Narrative Example
Due to low change success rates, high rates of change and high MTTR, IT is spending all of their time doing unplanned work.	"But the problem is, the development folks aren't the only guilty ones. Many of the operational changes that we make are just as risky. "We started to compute our change success rate, and while development only bats 40%, our change success rate is nothing to write home about either. In the last month, only 70% of the changes we made worked the first time without generating a firefighting episode."
Overall, there is a lack of confidence in IT.	"So, not only are we all ticked off with the performance, but we're also getting our butts kicked by senior management pretty hard these days. In fact, there are rumors that they are talking to potential outsourcers now, and that isn't doing morale any good."

Stabilize The Patient

> "To err is human. To really screw up requires the root password."—UNKNOWN

The first goal is to stabilize the patient. We need to decrease the amount of unplanned work in order to free up enough resources to create proactive processes. To do this, we start where the most damage is being done. Fortunately, this happens to be the place where we also have the most control. If 80% of our injuries are self-inflicted, then that means that we are causing 80% of our unplanned work. Therefore, we must start by reducing the number of self-inflicted problems by gaining control of the change process.

Start by identifying the systems and business processes that generate the greatest amount of firefighting. When problems are escalated to IT operations, which servers, networking devices, infrastructure or services are constantly being revisited each week? (Or worse, each day!) These items are your list of "most fragile patients" which are generating the most unplanned work. These are the patients that must be protected from uncontrolled changes, both to curb firefighting and to free up enough cycles to start building a safer and more controlled route for change.

For each fragile patient (i.e. server, networking device, asset, etc.), do the following:

1. **Reduce or Eliminate Access:** Clear everyone away from the asset unless they are formally authorized to make changes. Because these assets have low change success rates, we must reduce the number of times the dice are rolled.

2. **Document the New Change Policy:** Our recommended change policy is very simple: "Absolutely no changes to this asset unless authorized by me." This policy is our preventive control and creates an expectation of behavior.

3. **Notify Stakeholders:** After the initial change policy is established, notify all of the stakeholders about the new process. Make sure the entire staff sees it: email it to the team, print it out, and add it to login banners.

4. **Create Change Windows:** Work with stakeholders to identify periods of time when changes to production systems can be made. Our goal will be to eventually schedule all changes into these maintenance windows. Amend the change policy accordingly. For example, "Once I authorize the changes, I will schedule the work to be performed during one of the defined maintenance windows on either Saturday or Sunday between 3 and 5 pm."

5. **Reinforce the Process:** By now, you have defined a clear expectation of how changes are to be made. Make sure that people are aware of the new process and reinforce it constantly. For example, "Team, let me be clear on this: These processes are here to enable the success of the entire team, not just individuals. Anyone making a change without getting authorization undermines the success of the team, and we'll have to deal with that. At a minimum, you'll have to explain why you made your cowboy change to the entire team. If it keeps happening, you may get the day off, and eventually, it may prevent you from being a part of this team."

Electrify The Fence

> *"What is often overlooked is that if one person can single-handedly save the ship, that one person can probably single-handedly sink the ship, too."*—UNKNOWN

In the previous step, we have specified how and when changes can be made. This is the first preventive change process and policy. In reality, our experience has shown that merely specifying the correct way to make changes rarely results in everyone adhering to the process. We've found that managing change on the honor system is not enough, for a variety of reasons. Sometimes IT staff may be unwilling to change behaviors (e.g. "they're the IT cowboys who refuse to stop shooting from the hip"), development staff may not communicate well with IT operations (e.g. "they're the same developers who do not even show up to meetings"), or people may just make mistakes (e.g. "our best engineer made a seemingly trivial unauthorized change that just blew up").

So far, we have put a fence around the systems where unauthorized changes were causing the most carnage. In this step, we will electrify the fence. We do this to keep everyone accountable and responsible for playing by the rules. Our goal is to start creating a culture of change management. To do this, proper change monitoring must be in place so we can "trust, but verify." We will use this instrumentation to detect and verify that changes are happening within the specified change management process, and also to negatively reinforce and deter changes that are not.

We must be aware of changes on all infrastructure that we are managing: servers, routers, network devices, databases, and so forth. Each detected change must either map to authorized work, or it must be flagged for investigation. Critical questions that need to be answered are:

- Who made the change?
- What did they change?
- Should it be rolled back? If so, then how?
- How do we prevent it from happening again in the future?

Answering these questions forms one of the primary goals of the investigation process. Many organizations start the formal investigation by sending an email to the entire team describing the unauthorized change, and give the team a fixed time (e.g. four hours) for someone to step forward and explain why they circumvented the change process.

Although you can audit changes manually, change monitoring and reporting software, such as Tripwire®, automates the detection and reporting of changes. We recommend scanning systems for changes at least daily or after each maintenance window, whichever is more frequent. Almost universally, people implementing this phase are surprised and alarmed to see how many changes are being made "under the radar." Some organizations care so much about controlling unauthorized change that they have an end-of-shift audit process, where a change report is generated at the end of each shift, and the operations manager is required to attest that all changes can be mapped to authorized work or have been rolled back. This way, managers are held accountable for changes made on their watch!

Because we are now detecting changes that circumvent the change authorization process, we begin to "manage by fact" rather than "manage by belief." We no longer rely on verbal promises or assurances of good behavior. When a service fails, a server crashes, or an issue has been escalated to problem managers, we can generate and scrutinize reports of actual system changes. When stakeholders start seeing how often these issues are caused by failed change, a culture of change management starts to emerge.

The key to creating a successful culture of change management is accountability. If the change process is repeatedly bypassed, management must be willing to take appropriate disciplinary action, which may range from further training, public shaming, and eventually to formal HR-related measures.

One last comment on the importance of detective controls: Monitoring changes gives us a critical safety mechanism, just like a rock climber with a ratchet. The ratchet allows the rope to move in one direction, preventing the climber from falling. Monitoring change to enforce the process prevents our organization from sliding back into a state of uncontrolled change.

Modify First Response: The Catalytic Key

> *"Grant me the Serenity to accept the things I can not change, Courage to change the things I can, and Wisdom to know the difference."*—DR. REINHOLD NIEBUHR (EXCERPTED FROM THE SERENITY PRAYER)

In high-performing IT organizations, the change management process is catalytic, returning obvious and measurable value back to the organization daily. In this step, we will make our change process catalytic, to ensure that the organizations not only see the benefit, but also internalize and perpetuate the process. Despite all of our research findings to the contrary, change management is often viewed as a burdensome bureaucracy that consumes resources, time, money, and spiritual energy. To prevent this, we must replicate how the high-performing organizations use their change process. They integrate their change management processes with the problem resolution processes to drastically reduce MTTR. Specifically, when service outages occur, high-performing organizations first look at all approved and detected changes before making a diagnosis.

Why do they do this? Recall that 80% of all outages are caused by change, and that 80% of MTTR is spent trying to determine what changed. High-performing IT organizations eliminate change as a causal factor for an outage as early as possible in the repair cycle. They identify the assets directly involved in the service outage, and examine all changes made on those assets in the previous 72 hours. This information is then put into the work ticket, as well as the list of all authorized and scheduled changes. By doing this, when issues are escalated to problem managers, they have all relevant and causal evidence already at hand. Typically, when equipped in this way, problems managers can successfully diagnose issues without logging into any infrastructure over 50% of the time!

If no changes were authorized for the specific asset and no changes were detected, then the investigation circle is widened to the next ring of infrastructure that supports the affected asset. Again, we do this by examining our change management records and change monitoring systems, not by logging into infrastructure. For example, if the database service experiences an outage, we start our investigation by looking for authorized, scheduled and detected changes on the database server. If none are found, we then search for authorized, scheduled or detected changes on systems that support the database service, such as the operating systems, supporting networking devices, and the other dependencies. We have found that over 70% of service-affecting issues can be resolved in this manner.

Recall that the Microsoft Operations Framework (MOF) study showed that their best customers with the highest service levels rebooted their servers 20 times less often than average. This is because they manage problems by using causality and solving root causes, as opposed to "rebooting the server to see if the problem goes away." By integrating problem management with our change management processes, we will facilitate this same type of desirable behavior, and furthermore, change management becomes catalytic. IT operations will see value in knowing when changes are supposed to happen, and being aware of what changes actually occurred. In fact, many organizations have seen such value in this process that they complain loudly when this "instrumentation" is taken away—no one likes flying blind.

Helpful tips include:

- When creating the incident in the trouble-ticketing or change workflow system, pre-populate the ticket with all detected changes, as well as any other authorized changes made in the last 30, 60, and 90 days.

- If you do not find changes in the asset in question, increase the search radius to include the next ring of dependent infrastructure.

Create The Change Team

In the previous steps, we have started to specify the correct path for change and built the mechanisms to ensure that the process is being followed. In this next step, we will continue to develop the change management process by creating a Change Advisory Board (CAB), comprised of the relevant stakeholders of each critical IT service. These stakeholders are the people who can best make decisions about changes because of their understanding of the business goals, as well as technical and operational risks. Kurt Spence from HP states, "All business decisions result in an IT change event of some kind." Our goal is to make sure that they are fact-based decisions, resulting in managed changes.

Common stakeholders on the CAB often include the following people and roles:

- **VP of Operations**—is ultimately accountable for availability and has final authority on change approval
- **Director of Network Operations**—reviews priorities and impacts on resources
- **Security Lead**—reviews changes for security implications
- **Ops Systems Engineering Lead**—reviews changes for pre-production implications
- **Service Desk Manager**—reviews changes for customer-facing implications
- **Internal Audit**—may attend to better understand how changes are approved

One mistake organizations make is that they believe urgent changes (i.e. emergency changes) can be handled outside the CAB meetings. This assumption is false! Consider that virtually all cowboy organizations believe that their changes are both urgent and safe. In reality, emergency changes are the most critical to scrutinize and are the changes that require the most deliberation to approve. For these types of changes, create an emergency change process with a defined CAB emergency committee (CAB/EC) who can assemble quickly to review these requests. All changes that create risks must be evaluated and authorized, especially during emergencies.

Create A Change Request Tracking System

A prerequisite for any effective change management process is the ability to track requests for changes (RFCs) through the authorization, implementation, and verification processes. Paper-based manual systems quickly become impractical when the organization is large, or complex, or when the number of changes is high. Because of this, most groups use some computerized means to track RFCs and assign work order numbers. Some refer to these applications as "ticketing systems" or "change workflow systems." Examples of systems include HP Service Desk, Remedy ARS (Action Request System) and Best Practical RT/RTIR.

The primary goals of a change request tracking system are to document and track changes through their lifecycle and to automate the authorization process. Secondarily, the system can generate reports with metrics for later analysis. Each change requester should gather all the information the change manager needs to decide whether the change should be approved. In general, the more risky the proposed change is, the more information that is required. For instance, a business as usual (BAU) change, such as rebooting a server or rotating a log file, may require very little data and oversight prior to approval. On the other hand, a high-risk change such as applying a large and complex security patch on a critical production server may not only require good documentation of the proposed change, but also extensive testing before it can even be considered for authorized deployment.

Start Weekly Change Management Meetings (To Authorize Change) And Daily Change Briefings (To Announce Changes)

> *"If it is too complicated to understand, it is too complicated to govern."*—TOM HORTON

Now that we have identified the change stakeholders by creating the CAB, the next step is to create a forum for them to make decisions on requested changes. The CAB will authorize, deny, or negotiate a change with the requester. Authorized changes will be scheduled, implemented, and finally verified. The goal is to create a process that enables the highest successful change throughput for the organization with the least amount of bureaucracy possible. While they may seem unnatural at first, with practice, weekly 15 minute change management meetings are possible. Take special care to avoid an attitude of "just get it done," which allows people to make changes that circumvent the change approval process. If we make it easy for all changes to flow through our process, it will soon be easier to use the process than to circumvent it, even during emergencies.

CABs must meet on a regular published schedule that all stakeholders understand. To start, we will have each CAB meet weekly. The agenda will begin with recording attendance and progress to the following:

Deal With Old Business:

1. Review any failed changes or change management circumventions, as these are likely to have consequences.

2. Review and close action items from the previous meeting minutes.

3. Discuss any problems resulting from old changes. If necessary set up a Post-Implementation Review meeting to deal with issues.

4. Review any category of changes that should be categorized as "business as usual" to avoid repeated examination by the CAB.

Deal With New Business

1. Review the list of requests for change (RFC), and agree upon an order in which to evaluate them.

2. Examine RFCs for risks of collisions or interference, due to proximity of simultaneous or similar changes:

   a. Group changes by category (software, hardware, server, etc.).

   b. Examine proposed change dates for collisions and or potentially incompatible activities. (For example, multiple OS upgrades on the same day.)

3. Find any RFCs which are not actually tasks, but projects. Projects, unlike tasks, have multiple steps that are dependent upon each other. Verify that a project manager has been assigned for planning, coordination and execution. If the project dependencies have not been adequately evaluated, it is entirely appropriate to reject the RFC and request that new RFCs be submitted for each of the tasks.

4. Evaluate and authorize the RFC.

5. Schedule the approved changes on the Forward Schedule of Change (FSC) and assign a change implementer. (People with project management experience are great for coordinating these activities.)

6. Send rejected RFCs back to the respective change requester for further clarification or response to CAB comments.

The CAB meeting should be restricted to evaluating and approving RFCs, and should not get bogged down in process issues. Instead, dock these issues to be handled in a separate meeting. The goal is to keep the change management meetings focused on accomplishing the task at hand: management of change. When an RFC is rejected, the change requester should respond by addressing the concerns or providing more information. Remember that the function of the CAB is to identify which changes are risky, not to come up with solutions—doing this takes too much time. With practice, a CAB meeting can be finished in 15 minutes.

The following are typical questions to ask when evaluating a change for authorization. Not all changes require answers to all of these questions, but as the risk increases, insist on having good answers to more of these questions:

- "Who" Questions

  – Who will be affected by the change? Ensure that there is appropriate representation on the CAB to make decisions.

  – Who could be affected by the change if it fails?

  – Who from the potentially affected group(s) has signed off on the change?

  – Who is performing the change (the "change builder")?

  – Who has reviewed the proposed change?

  – Who is driving the change (the "change owner")?

  – Who is the project manager if this change involves more than one step?

- **"What" Questions**
  - What assets are the targets of the proposed change?
  - What is the change timeline?
  - What is the change review priority based on the associated risk and urgency?
    - Urgent—This change could cause a loss of service or severe impairment of usability to a large percentage of users or a mission critical business system and is needed right away. Immediate action is required and an urgent CAB or CAB/EC meeting may need to be scheduled.
    - High—This change could severely impact a large number of users. This change should be given the highest priority for change planning, building, testing and implementation in order to meet the next available maintenance window.
    - Medium—The impact of this change is not large, but can not be postponed until the next scheduled release or upgrade window.
    - Low—The change is important, but has relatively low risk and can occur during the next scheduled release or maintenance window.
  - What assets or processes depend on the targeted assets?
  - What will the successful change look like when implemented?
  - What business processes need to be verified after making the change?
  - What is the business or technical reason for the change?
  - What will happen if the change is not made?
- **"When" Questions**
  - When will the change be performed?
  - When will it be finished?
  - When will the benefits of the change be realized?
- **"How" Questions**
  - How will the change be implemented (in waves, one at a time, etc.)?
  - How will we verify success?
  - How will issues be escalated?
  - How successful were similar changes in the past? (i.e. change success rate)
- **"What if" Questions**
  - What is the rollback plan if the change should fail for some reason?
  - What is the worst possible outcome associated with this change?
  - What will the worst case service outage be?

Again, not all changes will require the same level of scrutiny. Business as usual (BAU) changes which are known, regularly executed, and have a low risk do not need such detail. Conversely, any change that is new or perceived as having material risk must have more detail to allow for accurate risk assessment by the CAB.

Miscellaneous Change Management Do's And Don'ts

> *"It's not the strongest species that survive, nor the most intelligent... but the one most responsive to change."*—CHARLES DARWIN

Here are some tips for change management.

Items to do:

- Do post-implementation reviews to determine whether the change succeeded or not

- Do track the change success rate

- Do use the change success rate to learn and avoid making historically risky changes

- Do make sure everyone attends the meetings, otherwise auditors have a good case that this is a non-functioning control

- Do categorize the disposition of all changes. In other words, all outcomes must be documented once a change is approved. Three potential outcomes are:

  - Change Withdrawn—the change requester rescinds the change request along with the reason why. This should not be flagged as a failed change in change metrics.

  - Aborted—the change failed, accompanied by documentation of what went wrong.

  - Completed Successfully—the change was implemented and is functioning appropriately.

Items not to do:

- Do not authorize changes without rollback plans that everybody reviews. Changes do fail, so be proactive and think ahead about how to recover from a problem rather than attempting to do so during the heat of firefighting.

- Do not allow "rubber stamping" approval of changes.

- Do not let any system changes off the hook—someone made it, so understand what caused it.

- Do not send mixed messages. Bear in mind that the first time the process is circumvented, incredible damage can be done to the process. "Well heck, we did it last time" or "The boss said, 'just do it'" both send the wrong messages.

Do not expect to be doing "closed loop" change management from the start. Awareness is better than being oblivious, and managed is better than unmanaged. Crawl, walk, run—when you put in a valve, you put it in the open position then you constrict as you have confidence that everything is flowing through it. The same is true for change management. Start with a particular class of changes and constantly refine the process.

The Spectrum Of Change:

The management of change is an evolutionary process. Groups should not become discouraged as they start developing their change management processes. The solutions may require changing people, processes, and technology. The following illustrates the stages of change management:

1. **Oblivious to Change**—"Hey, did the switch just reboot?"

2. **Aware of Change**—"Hey, who just rebooted the switch?"

3. **Announcing Change**—"Hey, I'm rebooting the switch. Let me know if that will cause a problem."

4. **Authorizing Change**—"Hey, I need to reboot the switch. Who needs to authorize this?"

5. **Scheduling Change**—"When is the next maintenance window—I'd like to reboot the switch then?"

6. **Verifying Change**—"Looking at the fault manager logs, I can see that the switch rebooted as scheduled."

7. **Managing Change**—"Let's schedule the switch reboot to week 45 so we can do the maintenance upgrade and reboot at the same time."

What You Have Built And What You Will Likely Hear

In this phase, our goal was to reduce the amount of time we are spending on unplanned work down to 25% or less, by reducing the number of self-inflicted problems and modifying how problems are solved so that change is ruled out early in the repair cycle. By increasing the change success rate and reducing MTTR, we have not only decreased the amount of unplanned work, but also increased the number of changes that can be successfully implemented by the organization.

This section of the book is not intended to be a complete reference document on change management processes. For further information, please refer to the chapter on change management in the ITIL Service Support volume. Copies of that book can be ordered from the ITPI Web site.

> "The first step is hard, but not nearly as hard as you might think, and the rewards are worth it. You are changing people's habits and ways of doing work, but don't forget that you are fundamentally making peoples' jobs easier as well. You are gaining control and creating a stable, maintained and predictable environment for people to work in. In my case, it took a security event to initiate this process, but we were able to remove developer access to production systems, remove unnecessary root privileges, and start down the path of building a functional change process. By doing the steps outlined in phase one of Visible Ops, we were able to reduce the catastrophic impact of change, and we bought ourselves 90 days during which we completed the rest of the Visible Ops phases. We involved the entire IT crew, who were very responsive and receptive, as were the other decision makers and stakeholders involved."—JOE JUDGE, FORMER INFORMATION SECURITY OFFICER (ISO), ADERO, INC.

"The value of the change management processes and detective controls is at this point incalculable to our organization. The old adage that "what gets inspected gets respected" applies here. As a result of increasing our awareness of change and the effects of change, we have changed the culture in regards to change. We are a more thoughtful organization; more conscious of change and more judicious in our use of change. The Change Advisory Board meetings have accomplished far more than the conscious planning of specific changes. They have raised the consciousness of change and the potential impact of change across the board. The fact that we are managing change at all creates this consciousness, which has a ripple effect throughout the organization.

It is not that we don't make mistakes anymore; but we have become more scientific in our approach to mistakes; mistakes are seen more as learning experiences and the mistakes have become fewer and farther between. The processes and detective controls have helped us realize many of our goals in the pursuit of world-class IT management."—STEVE DARBY, VP OF OPERATIONS, IP SERVICES

"The first phase of Visible Ops hits the nail right on the head with the focus on Change Management. In any IT Service Model that I am aware of, Change Management is at the very core. Starting with other processes (e.g. Help Desk/Incident Management) is merely staunching the bleeding, without addressing the underlying trauma."—JAN VROMANT, ITSM CONSULTANT

"We have seen tremendous value after implementing a change management process and detective controls. Unnecessary changes were identified and eliminated. This saved valuable time and moved us to a more proactive environment. Systems upgrades were reduced to mere hours instead of days as the processes were streamlined and standardization was easier to attain."—KAREN FRAGALE, DATA CENTER MANAGER

The benefits generated in this phase are:

- We have increased availability.
- We have reduced the time spent firefighting.
- We have increased the change success rate.
- We have created a formal change management process that is both documented and adhered to.
- We have reduced the risks of change that could negatively impact production.
- We have made failed change less costly and more visible by restricting changes to planned maintenance windows.
- We have reduced MTTR by ensuring that causal information is used by problem management, pulled from our change management processes.
- We have clearly defined the IT operations and security roles, welding them together in the change and problem management processes.
- We have improved the working relationship and communication between the functional roles. They are now working together to solve common business objectives, reducing the number of "drive-by" surprises.
- We have started creating a culture of change management, where the controls are owned by operations and security.

Helpful Tips When Preparing for an Audit:

- Avoid at all costs creating an adversarial relationship with auditors. Instead, demonstrate that you have effective management and control processes in place, and the documentation to prove it. If you cannot show intended and actual activities, auditors go into "archaeology" mode. (The worst thing you can do is become defensive and adversarial, especially if material control weaknesses do indeed exist.)

- Make sure you have an up-to-date document describing your change management process. Show this to auditors up front to illustrate what you want to be measured against. Without it, they will bring in their own processes to measure you.

- Take good meeting minutes during the CAB meetings and file them. Make sure they are dated. Show meeting minutes to auditors to demonstrate that the meetings are actually taking place.

- The mantra of post-Enron auditors is, "If it's not documented, it doesn't exist." Therefore, be sure to document both your work and your meetings. The correct level of documentation should be commensurate with the level of risk associated.

- To show that your change management processes function, meeting minutes should show:

  - Newly authorized and scheduled change requests.

  - Acceptance of implemented changes with correlation between detected changes and implemented changes, showing successful implementation, acceptance by a change manager and closure of the work order.

  - Changes to production equipment tracked in work logs/work order tickets. These should identify the date, time, implementer and system along with details of the changes made.

- Assemble a list of changes made outside of the change management policy and corrective actions taken.

- On a regular basis, create and review a report with the number of changes requested, changes approved, MTTR and Change Success Rate by asset, functional area and organization, etc.

- Engineer the change workflow and ticketing systems in such a way that "closing" a request or ticket is not possible until it has been reviewed and accepted by the change manager. This ensures accountability, visibility and fact-based management, instead of belief-based or faith-based management.

- By following these tips, you prove that you have functional preventive, detective and corrective controls in place. For more information, refer to Appendix A.

Phase Two: "Catch & Release" And "Find Fragile Artifacts" Projects

The second phase of Visible Ops focuses on creating and maintaining an inventory of production assets. Prior to the first phase, there was uncontrolled change and many different configurations in production. These configurations must be inventoried and analyzed to determine how to reduce configuration count, which will be the focus in Phase Three.

Issues And Indicators

The second phase of Visible Ops tackles the following issues:

Issue	Narrative Example
Inability to figure out where to start building a configuration management database (CMDB) and a service catalog that shows what services IT provides.	"Sometimes walking through the data center is a petrifying experience. There are hundreds of servers, which all look alike, that all seem to be generating firefighting situations. Which of these is really the most critical to our business? Is there a better way to find all the infrastructure dependencies than seeing them catch on fire when we make a change? "For that matter, what in the world do all these servers actually do? Sometimes when one of them goes belly-up, we get screamed at by someone we've never heard of. Normally, we wouldn't even listen to these people we don't know, but when they have 'VP' in their title, we sometimes can't help but scramble when they start yelling. How can we establish some prioritized list of services we're providing so we can do better triage?"
Inability to start moving from "individual knowledge" to "tribal knowledge." This is especially challenging when no documented processes exist that describe what the IT operations staff is responsible for.	"After weeks of cataloging all the IT services that we're responsible for, we finally have an up-to-date inventory. In fact, we also now have an asset inventory of the infrastructure that each service depends on. This still doesn't solve the problem that only a couple of rocket scientists understand how to run some of these assets. "For instance, take that DHCP server sitting over there. We know that if we turned it off, it would take down virtually all the middleware servers, but the engineer who set it up was a college intern four years ago who left after his internship ended. No one knows how to fix this thing. Virtually every attempt to modify this box results in a catastrophic failure and man-weeks of work trying to get everything up and running again! How do we make the DHCP server less fragile?"
Servers become like snow-flakes: Configuration drift creeps into mission-critical infrastructure, creating anomalous personalities in what should be identical infrastructure.	"Here's an interesting question for you. How in the world did that DHCP server even get there? Did you notice that the DHCP server is actually running on one of our four DNS servers? For seven years, each of those four DNS servers were exactly the same, until that college intern decided to commandeer one of them for his little project. "Obviously, when we first found that rogue DHCP server, we tried to kill it, but then found out during the middle of the trading day that all the middleware servers depended upon it. We never got around to getting rid of it, and the configuration variance is definitely starting to take a toll on other operational tasks. For instance, applying patches to all the DNS boxes is now starting to have radically different results because of the configuration variance!"

Implement A "Catch & Release" Project

"Insanity is doing the same thing over and over, and expecting a different result."—ATTRIBUTED TO
ALBERT EINSTEIN

By completing the first phase, we started to control how changes are made in production to reduce the likelihood of risky changes. In this phase, we will inventory all managed assets, and then identify those that create the most unplanned and unscheduled work. When we complete this phase, we will know exactly what infrastructure needs to be worked on by the release engineering team in the next phase.

In this step, we will do exactly what the park rangers for the National Wildlife Service do, which is "Catch & Release." Their job is to bag and tag each animal in the national parks, picking them up, weighing them, counting how many legs they have, giving them names, finding out if they already have a record, and so forth.

Our goal in this step is to capture all equipment in the data center. For each asset captured, figure out what it's running, what services depend upon it, who has primary management responsibility for it, how fragile this infrastructure is, and so forth. Attempting to find out this information in the heat of firefighting contributes to miscommunication and often results in tremendous stress and frantic phone calls. We will perform this inventory when infrastructure is not on fire, so that the information will be readily available in case of emergency. Note that the inventory is not only for use in problem management scenarios, but to guide resource deployment in Phases Three and Four, too.

While analyzing assets, ask these important questions:

- What does it do?
- What is the hardware platform?
- What is the operating system platform?
- What applications are installed?
- Who is responsible for this asset's uptime?
- What service(s) does it support?
- Who is authorized to make changes?
- What does this box do for the business?
- What will happen when this box stops working completely?
- What will happen when the performance of this box is severely degraded?
- What is the change success rate?
- Is this device fragile? Can we build a new one if it fails? (See the next Visible Ops Phase 3: Create a Repeatable Build Library.)
- What are its dependencies?
- What other infrastructure depend on this unit?
- What planned and unplanned changes have been made?
- What is the device's name? Is it appropriate for the tasks performed?

- What is the outage cost? (In other words, the cost per minute of downtime)
- Where is it physically located?
- Is there anything odd about this box?
- Is this a generally-supported platform in our company?
- Is this box going to go away in the next few months?
- How do we get access to this box (remote or otherwise)?
- How is this unit backed up?
- How long does it take to re-provision this unit (estimate)?
- How long can the business afford to be without it?
- Are we monitoring this unit for changes (i.e. is a detective control installed)?
- Are we fault-monitoring this device?
- Do the fault-monitoring assumptions match the dependency realities?
- If the unit is mission critical, then are there adequate hardware backups in place (power supply, network card, RAM, etc)?
- Why do we feel that this unit is unstable (if applicable)?
- Is there anything that needs attention on this unit?

Because collecting this information may seem like tedious data entry, you may be tempted to assign junior staff to this project. Do not fall into this trap! The goal of this step is to capture the information only known by the most senior staff, and use the information to create repeatable and verifiable processes. The senior staff, due to their experience and accumulated knowledge, will provide the bulk of this information that has not been previously captured. Currently, problems may escalate to senior staff simply because no documentation exists to allow more junior staff to resolve them.

Find Fragile Artifacts

"When you're in an earthquake on a unicycle, juggling chain saws, the only way to survive is to tack down everything you can tack down, so you can deal with what you can't."—STEPHEN CHAKWIN

Most data centers usually contain numerous infrastructure that are considered "dangerously fragile." During the "Catch and Release" process, these fragile artifacts must be tagged and treated as such. The senior staff usually knows where the fragile artifacts are because they instinctively shy away from them—knowing that if someone even looks at one wrong, it will crash and cause a massive episode of unplanned work. Infrastructure is fragile when it has a low change success rate and a high MTTR. In other words, fragility occurs when all changes are risky and potentially require a rebuild from scratch, resulting in several man-weeks of work to repair.

Once we have a list of fragile infrastructure, what do we do with it? First and foremost, avoid making any changes to them! Consider how valuable it was to avert risky changes during the change management meetings. By flagging fragile infrastructure that generates inordinate amounts of unplanned work, we can further avoid risky changes. During this step, some groups have literally put Post-It notes on infrastructure boldly warning "Do Not Touch!"[9] Others use login banner screens or announcements to convey the same message. And of course, these are the systems that need detective controls in place, to ensure that the change management process is not circumvented!

Beyond the value of deterring changes, the "Do Not Touch" sign has further value because it can allow us to make better decisions in the CAB meetings. What is the value of preventing a change that would have resulted in 200 man-hours of unplanned and unscheduled work? There is a well-documented story of a $239 million weather satellite that was accidentally dropped because someone had removed the bolts that secured it on the stand, and did not bother to put a sign up warning people not to touch the delicate satellite![10]

Throughout all four phases of Visible Ops, we foster and positively reinforce organizational learning. Einstein once defined insanity as "doing the same thing over and over, and expecting different results." Here, we document what works and what does not work to guide decision making, and to avoid actions that historically generated unplanned and unscheduled work. We will leverage these findings in the third phase to increase the number of planned, tested, stable, and repeatable configurations.

Prevent Further Configuration Mutation

"Do not look where you fell, but where you slipped."—AFRICAN PROVERB

Uncontrolled changes on systems cause them to deviate from any known and trusted states. We may deploy one thousand servers that are initially identical, but when their configurations drift over time, they become like

---

[9] Bill Shinn tells the story of a former Dutch boss who would walk around the server room placing "Do Never Touch" Post-It notes on the critical servers.

[10] Fordahl, Matthew. "Under-construction satellite topples to floor in mishap." The Associated Press, 9/10/2003.

snowflakes—each different and impossible to reproduce. It is absolutely critical that while we inventory systems in this entire "Catch & Release" phase, they must not mutate into some other unrecognizable configuration. (Recall the story of the DNS server suddenly becoming a DHCP server as well.)

During Inventory Projects:

While we are doing the "Catch & Release" inventory, we must freeze configurations. If new builds are silently deployed into production or previously inventoried builds change, we jeopardize the accuracy of our inventory. Therefore, we must do the following:

1. Create a clear mandate delaying deployment of new builds and changes to existing builds until the inventory process is over. Set clear goals for dealing with each fragile patient, defining the start and finish of the change-free moratorium, and communicating with business units to define a mutually acceptable change freeze window. The last thing we want to do is unnecessarily delay change and prevent a new product rollout—that's as bad as an unplanned outage from the perspective of a business unit.

2. The "Catch and Release" and "Find Fragile Artifacts" steps must happen quickly. It is unrealistic to prohibit changes for too long. This may force people to circumvent the change process in order to get critical work done.

3. Allow truly urgent and necessary changes to be made through the CAB/EC meetings, making sure to capture any changes to systems that are only partially completed with their Catch and Release treatment.

During Ongoing Operations After Completion Of "Catch and Release":

An automated detective control must be used as a ratchet to ensure that any progress made is not immediately lost due to entropy. This control must detect all changes, notify us when processes have been circumvented, and enable an update to the configuration inventory. Detected changes that cannot be mapped to authorized work must be rolled back.

What You Have Built And What You Will Likely Hear

> "Every time a security or IT organization has done an asset/service inventory I've seen only two types of reactions. On the mild side, you have 'That's odd. Where'd those come from?' On the stronger side, which is more typical, you have 'Houston, we have a problem!'
>
> Seek and ye shall find—what you have never matches what you think you have. The best organizations merely keep that gap small, safe and manageable."—JOE JUDGE, FORMER ISO, ADERO, INC.

> "We found that creating and managing asset and service inventories is essential to achieving change management. Our asset inventory, in the form of a Configuration Management Database, gives us a reference point as to what the configuration of the IT infrastructure should be as well as a record of changes to the infrastructure. The CMDB supplies a contextual environment for managing change that enables management of change from the perspectives of relationships between entities as well as history. The CMDB quickly became an essential reference for change management. It is hard to imagine managing change without an accurate CMDB.

In a way, the service inventory fills a role that is similar to the asset inventory. It is another reference, this time describing the policies and procedures that govern operational activities that result in changes being made to the configuration of assets. Solid, detailed services documentation is another essential element for successful service management. By documenting the services in detail, including how the services are to be implemented, we have improved our control over service management. Failed changes can usually be traced back to someone not following documented procedure or an incorrect or incomplete description of a procedure. By documenting known good procedures and ensuring that staff is trained to follow the procedures, we have a powerful system for increasing our change success rate.

Services and assets can be seen as the inputs and outputs of the change management process. By documenting the types of services that are authorized to be performed and documenting the status of the IT infrastructure upon which they are performed we have completed the system that is required to track change scientifically and generate management feedback in a context that enables continuous process improvement."—STEVE DARBY, VP OF OPERATIONS, IP SERVICES

"The identification of the most fragile items in the IT architecture is, in my humble opinion, a crucial second step, and I was excited by the fact that Visible Ops recognizes it as such. I once heard a well-known consultant say 'configuration management has to be done on the sly, as you can not get any wins from there.' Nothing could be further from the truth, and the second phase of Visible Ops proves it in a logical way! Configuration management and change management are two sides of the same coin, and one cannot work without the other. Phase Two of Visible Ops focuses on the core configuration management—right on target!"—JAN VROMANT, ITSM CONSULTANT

"It has always been important to keep an asset inventory. It is also extremely important to inventory your services and processes. It has helped move our organization from a reactive environment to a proactive environment."—KAREN FRAGALE, DATA CENTER MANAGER

The benefits generated in this phase are:

- We have created a service catalog that documents the most critical services we are supporting.

- We have documented what those services are and what infrastructure supports them.

- We have created a configuration management database (CMDB) that illustrates the mapping between services and infrastructure, showing the relationships between all the configuration items (CI). Each CI is associated with a service and may have other CIs associated with it. An example CMDB table structure is provided in Appendix E.

- We have created decision support tools and metrics that increase change success rates and further decrease unplanned work.

- We have fostered organizational learning by using historical change success information to make better risk-driven decisions around change.

- We have created a prioritized list of projects that our release engineering team will work on, to replace fragile artifacts with stable infrastructure.

- We have implemented detective controls to ensure our inventory information is always up to date. The detective controls now are being used for both problem management and ensuring the integrity of the CMDB.

Helpful Tips: When Preparing for an Audit

- Be able to show the process of how you generated the catalog of services and the assets that support them. Remember, the inventory should include both hardware and software.

- Show that you understand the business processes you are supporting by working with senior management to rank the services by importance to the organization and their degree of fragility.

- Show how you assure that the inventory is maintained and accurate. The lack of an accurate inventory may indicate to auditors that there are inadequate controls.

- Show the list of fragile artifacts resulting from this phase as evidence that you are performing risk mitigation.

- Be able to document the systems and processes used to detect changes.

Phase Three: Create A Repeatable Build Library

In phase two, we inventoried all of the IT services and the dependent infrastructure. We identified fragile artifacts with low change success rates and high MTTR because of their contribution to unplanned work. Then we plastered them with Post-It notes warning "Do Not Touch!"

In this phase, we will create a library of repeatable builds focusing first on fragile infrastructure. To accomplish this, we must define build mechanisms, create system images, and establish documentation. What results is a repeatable process for building infrastructure from "bare metal."

By making infrastructure easier to rebuild than repair, we will create the data center equivalents of fuses: when things go wrong with production fuses, we do not repair them; instead, we pop them out and re-provision them from scratch. However, success in this phase hinges on our ability to control production changes to that fuse, lest the new fuse have radically different behavior than the fuse it replaces. This technique replicates what all the high-performing organizations do, resulting in the high server/admin ratios, low amounts of unplanned work, and the ability to maintain manageable configuration counts.

Issues And Indicators

Issues	Narrative Example
Critical production infrastructure resembles works of art, which can never be replaced, even during disaster recovery situations.	"Remember that supposedly simple 'DHCP server' I was talking about? Let me tell you about our last attempt to replace it. In fact, I have an email from the build engineer that wrote me during our last disaster recovery test last spring. It reads:
	Dear, Bob. I am writing this to you from the data center that supports the billing middleware. I do not need to remind you that we haven't turned this stuff off in over four years. We are going to be power-cycling it in ten minutes. No one on the team, myself included, knows whether we will survive the reboot. Worse yet, if it doesn't fully boot up on its own, we will have no idea what we need to do to bring it back up. That's why the entire staff is ready to spend the weekend on site.
	Wish me luck.
	Ellen
	"Well, 72 hours later, we got it up, but just barely. If Ellen hadn't been able to hunt down people who managed this stuff four years ago, I don't think we could have pulled it off. Talk about archaeology—what would have happened if one of those people were no longer alive? What if we didn't have access to Google? How in the world can I make sure this never happens to me again?"
Configuration drift undermines release management investments. In the absence of "refrigeration," infrastructure "spoils" in production.	"So what is the point of having our release engineers work on repeatable builds for the DNS servers, when the instant they go into production, they get patched, changed, configured, re-configured, over and over again. Meanwhile, the release engineers are working on their next release, spending weeks getting it loaded up into Microsoft SMS or Marimba, blasting it out and then overwriting all the production changes we made! Suddenly, we in IT operations have to reapply all the changes we made in the last two months. Heaven help us if we didn't document all of our changes."
	"I probably mentioned that we sometimes want to strangle all the developers because of all the junk they throw over the wall at us. Unfortunately, equally often these days, we want to strangle the release engineers as well."

Issues	Narrative Example
Inability to decide where to apply release engineering resources: what builds should they be working on, and when is it good enough to move on?	"Because of all the configuration variance, the release engineers have a difficult time determining how to reduce configuration complexity. Even worse, the production environment is a moving target, because it changes so quickly. "Although release engineering tries to keep up with the production environment, they rarely can, due to all of our patches and modifications. Inevitably, we have to spend a lot of time rebuilding their boxes and doing testing until we hope we've caught all the possible errors. As you can imagine, we definitely miss problems once in a while and they pop up during the worst times!"
Downward spiral of increasing configuration counts, and increased level of knowledge necessary to resolve problems, which increases the need for more highly skilled system administrators.	"Quite frankly, I'm starting to really understand that I cannot avoid being in a downward spiral. My production configurations continue to turn into irreplaceable snowflakes, which require an increasing level of skill to make any changes and administer. In turn, this is increasing my need for rocket scientists, forcing me to hire ever smarter people to do more non-repeatable tasks. If I keep this up, then I'll have one rocket scientist for each server, who'll be sitting on the bench doing nothing most of the time! "In reality, I need to be going the other way. I need to get my rocket scientists out of the front lines and building tools to empower the production staff to diagnose and fix the problems!"
Development routinely makes fixes on production infra-structure instead of providing tools for Ops staff. They "throw the pig over the wall" to Ops with little ability to control, stage, or reject.	"Humbly speaking, I feel that I have enough problems with the rocket scientists in the development organization. I should be creating a process where application developers are better integrated with operations. Somebody needs to take the time to package them for distribution and installation in a way that doesn't require hundreds of hours of integration engineering, and that can be accomplished by production staff! "For this to work, the production staff needs to have the ability to say 'no' on production changes. If pushing out a release has a 25% failure rate, then we need to be able to push it back to development to rework. What use is deploying software if it will cause huge amounts of unplanned work and takes down revenue-generating infrastructure?"
Ops routinely crashes infrastructure by applying security patches.	"I sat in a SANS class on applying the 'Gold Standard' configuration to Windows servers. Definitely all smart things to do, but do you want to talk about risk? There were 130 people taking the class, each person bringing their own laptop to learn how to apply the changes. By the end of the six hours, five people had laptops that could no longer boot! "It finally dawned on me. Security is often recommending 'fixes' that have a 3–5% failure rate, which is the reason why my heart stops whenever I see one of these guys submit an 'emergency change request' to apply these security patches!"
Ops applies infrastructure patches that never get reflected in future builds.	"This makes it all the more important that if we go through the agony of figuring out how to successfully deploy these security patches, they absolutely must make it into the stored golden builds. Otherwise, we're going to have to jeopardize the infrastructure and spend hundreds of man-hours redoing all the work!"
The production team supports unusual applications and infrastructure.	"How in the world did we get into the position where we are supporting fourteen different database applications? I'm not even sure I know what a 'Kumquat Express Postgres IV' is, but it catches on fire a lot, we know how to reboot it, but that's about it. I think it was some marketing person who bought it on their credit card, put a pig around it, and threw it over the wall at us."

Create A Release Management Team

In this phase, our goal is to take the most senior IT operations staff and move them from a reactive firefighting role to a proactive release management function, where they are constantly working on software and integration releases that will get deployed into production. By doing this, they operate early in the IT operations lifecycle where the defect costs are lowest. Most likely, the team was also involved with the previous "Catch & Release" phase and is very aware of the systems that are in production.

The release management team's primary role is building the mechanisms to deploy the best configurations into production. They do not do the builds—they engineer the builds. In other words, they design the builds, but don't build the servers. In this phase, their first task is to create repeatable builds for the most critical and fragile infrastructure assets identified during the "Catch & Release" phase. Their objective is to make it cheaper to rebuild infrastructure than to repair. Ideally, this will be done by taking the affected infrastructure, burning it down to bare metal, and rebuilding it by pushing a button.

In increasing order of importance, the following typical benefits result from rebuilding instead of repairing:

• Rebuilding infrastructure is an automated process and takes a known amount of time, as opposed to firefighting and repairing, which almost always take longer than we originally estimate.

• Rebuilding infrastructure tends to introduce less configuration variance, as opposed to repeated break/fix cycles which allows additional configuration variance to creep in.

• Because the rebuild process is automated, documented and less complicated, it can be done by junior staff, freeing up senior staff from firefighting.

• When senior staff become free from firefighting, they can keep working on new build projects, which will fix other systemic issues.

The release management team continually works on proactive projects that reduce the likelihood of unplanned work, systematically eliminating the sources of disasters before they strike. They strive to reduce the number of unique configurations in production and increase the lifespan of a configuration before it needs to be changed or replaced. Both of these goals reduce complexity and cost while improving manageability.

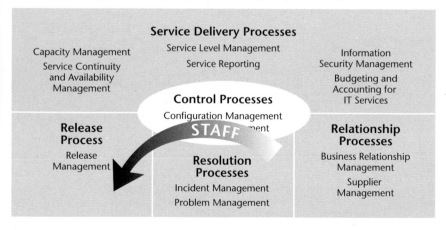

Figure 5: Staff deployment gradually shifts from reactive resolution processes to proactive release processes

By moving the most senior staff to the release engineering processes, we better equip the more junior staff to maintain production infrastructure. By allocating staff this way, the organization uses people more effectively at all levels: Their mastery of configurations continually increases while they integrate it into documented and repeatable processes. We jokingly refer to this phenomenon as "turning firefighters into curators," because of the tendency for the best and the brightest engineers to be the worst instigators of self-motivated firefighting. By putting them into the build process, they become protective of post-deployment build integrity, and actively discourage production firefighting! At the same time, more junior levels of the organization are able to take over problem resolution.

In the following steps, we will create a process for the release management team to generate repeatable builds, store them in the definitive software library (DSL), and put together maintenance and update plans for the builds. We will also create a process for the operations team to take the builds from the DSL, and then provision them into production.

We will describe the DSL in more detail in the next several steps.

Create A Repeatable Build Process

Now that we have created a release management team, they need some projects to work on. Not coincidentally, this is the output of the "Find Fragile Artifacts" step of the previous phase. We have already created a prioritized list of fragile infrastructure that needs to be replaced with stable infrastructure that can be repeatedly rebuilt by junior staff.

Golden builds are the output of the repeatable build processes. They are considered "golden" because they have been through the planning, testing and approval processes prior to being pushed into production. Builds must be updated when new patches and upgrades are integrated and new applications added.

The golden builds are stored in the DSL. Think of the DSL as the vault where all required software assets reside. The contents of the DSL are used to construct and reconstruct production infrastructure. It includes software media, license keys, software patches, and so forth. The DSL is the authoritative and secure storehouse of all software that has been reviewed and approved for production use.

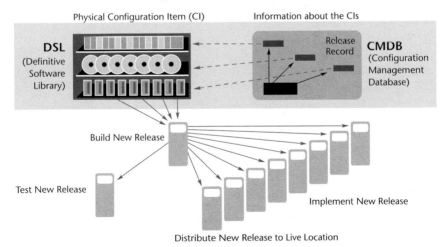

Figure 6: ITIL "Definitive Software Library" Diagram

To define which build projects the release management team should be working on, do the following:

1. Identify the common set of essential services and components used across your infrastructure. These are the lowest common denominators that apply everywhere. To start, consider creating lists of supported components for the following categories:

   a. Infrastructure and operating systems

   b. Applications

   c. Business rules

   d. Data

2. From these components, create a list of standardized components, called a "build catalog." Look for ways to create components that can be reused and combined to create standardized configurations. For example, a Web server build may be composed of a Solaris build, with the Apache build installed over it. The database server build may be the same Solaris build, but with the Oracle package installed over it.

3. For each component in the build catalog, create a repeatable build process that generates it. The goal is to have an automated build system that can provision the package by "pushing a button." Examples: For AIX, use NIM; for Solaris, use Jumpstart; for Windows, use InstallShield AdminStudio to "diff" systems and create HAL-proof installers.

4. Any testing or lab environment should be isolated from the production network to ensure that it does not disrupt production systems and to make sure that all dependencies outside the test environment are fully documented and understood.

5. Ensure that you can recreate each system with a bare metal build. Our goal is a repeatable process that eliminates anything tedious and error-prone, as well as reducing labor, errors, and the amount of spiritual energy required to maintain "snowflake infrastructure." (Note that VMware is a useful tool to create identical virtual servers for integration testing.)

6. For critical high-availability or load-balanced environments where many machines perform the exact same function, develop reference builds that can provision a box from bare metal without human intervention.

7. When the build engineering process has been completed, store them in the DSL, making the build available to the provisioning teams. Creating and maintaining the DSL is covered in the next step.

Create And Maintain The Definitive Software Library (DSL)

Ideally, the release management team will not be doing the actual provisioning and deployment of infrastructure into production. Instead, the projects they complete will generate builds that are put into the DSL. These are then used by the operations team to deploy the various builds into production.

In this step, we will describe the processes for how we will create, populate and maintain the DSL:

Generate The DSL Approval Process

1. Designate a manager to maintain the DSL, who will be responsible for authorizing the acceptance of new applications and packages. Of course, in the beginning, the DSL will be empty, which brings us to the next step.

2. Create an approval process for accepting items into the DSL. For example, we want to prevent some business manager demanding the use of a new IIS 6.0 version to support some new marketing Web site, which then requires a whole new production skill set to maintain. Ideally, acceptance should require approval from both the operations and release management staff, as well as a relevant expert. Furthermore, the prospective builds should pass a pilot or lab trial prior to inclusion into the DSL.

3. Establish a provisioning vault or clean room to store approved media for the DSL and to build servers. Its network should be isolated and not directly connected to the Internet or other IT networks.

4. Any software accepted into the DSL (both retail and internally developed applications) must be under revision control.

5. Audit the DSL to ensure that it contains only authorized components.

Place Repeatable Builds Into The DSL

1. Initially, put all the running applications into the DSL, under a special "amnesty program." Because we do not have repeatable builds for them yet, tag these as having amnesty for a fixed period, e.g. one year.

2. For each application in the amnesty program, create a repeatable build package, using as many of the generated builds that were previously defined. Our goal is to replace each program under amnesty with as many pre-defined components as possible, to best leverage economies of scale. Benefits include the ability to do mass upgrades, reduced configuration variance, and so on. For example, we may initially have 14 different Solaris and Windows builds, but our goal will be to engineer[3] builds to replace them all.

3. Periodically review the DSL to weed out packages that are no longer useful, maintainable, or cost effective to keep around. For example, after reviewing requirements, we find that instead of five database packages, the same jobs can be handled by just three.

Create An Acceptance Process Contract

Often, unclear roles for the release management and production teams create a situation where they undermine each other, instead of working together to solve common objectives. To avoid this, create clear organizational responsibilities and shared goals for both teams, and define the interface and working relationship between them. The primary mechanism to accomplish this is the acceptance process, where the production organization decides whether or not to accept what the release management team has built.

A successful acceptance process allows both teams to solve the common business problems, despite their slightly different objectives. The production team should focus on their desire for platform stability, ease of manageability for production staff, predictability, and fully tested functional products (as opposed to "do-it-yourself" kits). On the other hand, the release management teams should optimize for delivering against clearly defined requirements, ability to have longer release cycles, which depends on being free from production firefighting.

To define the acceptance contract, have the most senior persons from both the production and release management teams define how pre-production builds should be deployed into production. Critical questions include:

• Who designs and specifies the production environment requirements?

• At what point does operations get involved?

• What is the nature of the hand-off between release management and production?

- Who is responsible for creating build criteria and scheduling build deployments?

- What are the build dependencies? For example, for Java Virtual Machines (JVM), which versions and vendors will be officially supported? How will unsupported JVMs be transitioned to supported JVMs?

- What hardware is on production and development systems? Can the build systems support disparate systems? (For example, certain lower cost Intel server platforms are notorious for switching components weekly, undermining the ability to have uniform builds due to driver requirements.)

- What are the roles of machines used by release management? (For example: test, staging, production, developer toys, etc.)

- Will release engineers have access to production systems? (Incidentally, this is strongly discouraged and is discussed in the next section.)

- Do all release plans have a valid back-out plan before being accepted for deployment into production? (Of course, the CAB will oversee and enforce this as well.)

Moving From Production Acceptance To Deployment

Remember, the release engineers design the builds. Operations staff must accept and then use the build process to provision infrastructure into production. Provisioning infrastructure into production will involve the following steps:

1. The release management team must first have checked in all of the necessary build tools, software media and documentation to the DSL. Production staff will check out the builds from the DSL, which will be used to build and deploy the production infrastructure.

2. Remember that the team responsible for building new infrastructure has separation of duty requirements. For example, no developers should be part of the build process, both for security reasons and to advance the goal of the operations staff's ability to provision and maintain infrastructure unaided.

3. The operations team will provision the system. Optionally, a QA team may first test that the system is actually stable and functional, or it may be deployed into production on a limited basis on low-impact and non-essential services, or a combination of the two.

4. The operations staff will submit an RFC into the CAB to get approval and a schedule for system rollout.

5. Once the deployment is approved and scheduled, it is released into production.

Define Production Plan For Patching And Release Refresh

Despite the urgency attached to applying software patches, patch deployment ideally belongs in the release management process. The "patch and pray" phenomenon is well-documented; it refers to the fact that neither patching nor avoiding patching seems to achieve the objectives of creating an available and secure computing platform. We observe that high-performing IT operations organizations patch far less frequently than typical IT organizations, and yet they still achieve their desired security posture. It is incorrect to assume that they do this at the expense of security! Rather, they effectively manage risk and use compensating controls instead of patching.

Strive to apply and test patches on the pre-production systems before deploying changes into production. Ensure that the pre-production and production systems stay in sync to the extent possible. While production and pre-production hardware may be radically different (e.g. high-end vs. low-end platforms), configuration variance can be managed so that testing and qualifying changes (including patches) can be effective. Creating repeatable builds for test systems allows for the constant "blowing up of boxes" to test new configurations, while guaranteeing that a clean copy of the pre-production system exists that is 100% in synch with that in production.[11]

Due to the complexity of modern systems, patching in production can easily create errors, both immediate and latent. Latent errors can accumulate, increase configuration variance, create system errors, and may even compromise security. By moving patching into the pre-production test environment and the release management process, you are more likely to catch errors due to the better control and testing.

Ideally, we want to create a process where systems are being provisioned into production in a planned and scheduled manner instead of always being deployed in an urgent and ad hoc manner. To do this, we will create a process that evaluates the patches and issues from problem management, as well as security and update bulletins from subject matter experts (e.g. CERT, SEI, SANS, vendors, etc.) for applications in the DSL.

Consider evaluation questions such as:

• Is this a material threat to our ability to deliver safe and reliable service to the business?

• Can we mitigate this threat without applying the patch or update?

• Can we test the impact of the update and feel confident that our tests will predict the outcome on our production systems?

• When is the next release cycle? Can we package this update with other tested updates?

• If we have to do this now, how can we minimize the risk of unexpected consequences?

• If we cannot reduce the risk of exposure through testing, and we cannot bundle this with any other releases, then can we get the stakeholders and IT management to sign off on the risk?

• Create a release schedule that achieves the above objectives, attempting to bundle patches and updates into releases instead of applying individual patches to machines. Obviously, each of these releases needs to involve the CAB for review and approval prior to deployment. We want to keep the release management team focused on longer-lifecycle projects, which get deployed on a regular basis. The longer the shelf life of the production builds, the more stable the production infrastructure will be.

Close The Loop Between Production And Pre-Production

Before we entered this phase, we started to manage the change process to avert risky changes. Now, as we bring the release engineering capabilities to bear on the problem, managing and documenting changes become even more important. In order to maximize the post-deployment configuration integrity and shelf-life of the production builds, all production changes must be reflected in the new builds, lest they be overwritten by a new software rollout. This is to ensure that when infrastructure is replaced or rebuilt, it is replaced by systems that are functionally identical to the original systems. If you are treating production infrastructure like fuses, success hinges on the production and stored fuses being identical.

One step you can take, which is actually required in most regulated industries, is to prevent developers and release management engineers from accessing production infrastructure. The notion that "those who built the

<hr>

[11] The NIST has a reference document about security patching at: http://csrc.nist.gov/publications/nistpubs/800-40/sp800-40.pdf.

airplane are not allowed to fly it" is called "separation of duty"[12] and is required for security reasons to ensure that no single person can introduce uncontrolled production changes. For the same reason, we recommend separating duties in IT despite the cultural difficulties it poses (i.e. it is often difficult to take away root access from developers, but we have found that having an availability level greater than 95% is difficult without having clear separation of roles).

To reiterate, to be certain that production builds stay in synch with the golden builds, use a detective control mechanism to ensure build integrity. These days, systems have tens of thousands of files, and hundreds, or even thousands, of configuration options, and countless file versions. Without some automated way of scanning for changes, it is far too easy to make a change (accidentally or not) and then not record it. Only through the automated detection of changes can we close the loop between production and pre-production.

What You Have Built And What You Will Likely Hear

Successfully implementing this phase will typically reduce the amount of unplanned work down to 15% or less. By drastically reducing configuration counts, we can significantly change the staffing allocation from unplanned to planned work, and consequently increase the server/sysadmin ratio.

> "The single, largest improvement an IT organization can benefit from is implementing repeatable system builds. Being a systems administrator at heart, I cannot stress how much this will positively influence any systems management organization. But you can't do this without first managing change and having an accurate inventory. When you convert a person-centric and heavily manual process to a quick and repeatable mechanism, the reaction is almost always very positive.
>
> Using a defined image for a Jumpstart, Kickstart, disk clone, Ghost, or any other automated installation system will immediately decrease the error-prone, manually intensive system setup and configuration process. Even partially automated release/build processes greatly improve the ability for individuals and organizations to be freed from firefighting and to focus on their areas of real value and expertise. And by making it more efficient to rebuild than repair, you also get much faster systems recovery and significantly reduced downtime."—JOE JUDGE, FORMER ISO, ADERO, INC.

> "Before the application of change management processes and controls on our release management processes, we lacked optimal control over individual instances of production systems. Each system was at some level unique and would behave in unique ways. Recreating systems to exactly the same specifications was not always possible. Now, we are able to record every detail of a build with 100% certainty and ensure that production systems remain identical to our golden builds. This enables us to rapidly remediate problems with confidence. This is yet an additional benefit that comes from change management. Good change management results in achieving the ability to maintain known good configurations which can be applied to remediate failed changes or other system failures."—STEVE DARBY, VP OF OPERATIONS, IP SERVICES, INC.

> "We can identify changes immediately and match them up with documented planned changes. We have documentation, testing procedures and back out plans required in the change management process. This allows us to quickly rebuild and identify what changed and the impact it had to our systems. Time is a precious commodity and with our tools we our managing our time more effectively and efficiently."—KAREN FRAGALE, DATA CENTER MANAGER

---

[12] Also known as "segregation of duties".

Benefits arising from this phase are:

- We have created a release engineering team with well-defined roles and responsibilities, to enable an effective working relationship with the production team.

- We have created a process for the release engineering team to define and generate infrastructure that can be repeatedly built.

- We have further decreased uncontrolled production changes, which increases the amount of time available to work on planned tasks.

- We have created a new problem resolution mechanism, making it cheaper to rebuild than to repair. This creates an alternative to protracted fire-fighting.

- We have enabled shifting of senior staff from the front-lines to the release management area, where the defect repair costs are lowest. This further enables junior staff to handle ever more challenging production responsibilities.

- We have moved more staff off the front lines to work earlier in the lifecycle.

- We have closed the loop between production and release management, to curb production configuration variance.

- We have enabled the continual reduction of unique configurations in deployment, increasing the mastery of each configuration in deployment, increasing the server/sysadmin ratio.

- We have mitigated the "patch and pray" dilemma, by integrating software updates into the release management processes, where patches can be safely tested and rolled out.

Audit Tips:

- Fully document the build process from feature request, to build definition, to build acceptance.

- Fully document the acceptance and handoff process between the pre-production and production teams.

- Prepare reports on production rollouts of software, change success rate, time required to complete the rollout, and the integration with the change management processes.

- Document the process of how software is evaluated, accepted into, and purged out of the DSL.

- Generate a report of the percentage of deployed systems that match the golden builds.

- Document the process used to track threats and generate projects in the release management processes for patch updates and software rollouts.

- Document the policies for the clean-room build process.

- Be able to show how systems are certified. In other words, "How do I know that what I built is what I intended to build?"

- Be able to provide a list of all exceptions to the golden builds and justifications for them. An abundance of unexplained exceptions is evidence of an ineffective process.

Phase Four: Continual Improvement

"A vision without a task is but a dream. A task without a vision is but a drudgery; but, a vision and a task are the hope of the world."—UNKNOWN (FOUND ON WALL OF CHURCH OUTSIDE SUSSEX ENGLAND, CIRCA 1700)

Philipp M. Nattermann wrote in the McKinsey Quarterly that if all you are doing is adopting best practices, then eventually, all you are going to get is competitive parity.[13] In order to really excel, you need to optimally apply all your resources to achieve the real business goals. To do this, we want to make sure that we are best applying resources towards the accomplishment of objectives. In a world where competitive business and security pressures keep increasing, the only way to effectively respond is to stabilize the environment and reduce firefighting. This results in more time available to address the needs of the organization. One of the best ways to manage and allocate resources during this continuous improvement journey is with metrics.

Metrics And How To Use Them

"The better you get, the better you'd better get."—DAVID ALLEN

A management truism is that you cannot manage what you cannot measure. In the case of IT and automated systems, it is far too easy to start collecting a mountain of data and generate metrics that are of little or no value, regardless of how much you analyze them. In this phase, we will focus on metrics that aid decision making and give indicators as to the qualitative type of work being done (i.e. planned vs. unplanned, active vs. reactive, early vs. late in the lifecycle, and so on). We will discuss the types of metrics we can generate, and describe how to use them to guide improvement efforts.

In general, the key metrics for IT operations are the availability metrics, such as the ITIL resolution process metrics of MTTR and MTBF. The problem is that virtually all the factors that affect these metrics live in the controls and release process areas. Essentially, think of MTTR and MTBF as being symptomatic of decisions made elsewhere.

Upon completion of the first three Visible Ops phases, we have created fully-functional release, controls and resolution processes as well as having created a closed-loop feedback mechanism between all of the three process areas. In fact, we have done more than that: We have created a minimal closed-loop system that is capable of improving itself. What do we mean by this? By completing the previous three phases, we can now generate metrics for the three key process areas (release, controls and resolution) that dictate the following:

- **Release**—How efficiently and effectively do we generate and provision infrastructure?

- **Controls**—How effectively do we make good change decisions that keep production infrastructure available, predictable and secure?

- **Resolution**—When things go wrong, how effectively do we diagnose and resolve issues?

---

[13] Nattermann, Philipp M. McKinsey Quarterly 2000. Reprinted in BetterManagement.com
http://www.bettermanagement.com/Library/Library.aspx?a=11&LibraryID=8387

We can generate metrics from these three process areas, and use them to guide investment in future improvement efforts. After all, the goal of the first three phases was to stabilize the environment and free up enough resources to work on proactive projects that deal with root causes. With those phases complete, we have laid the groundwork to focus on continuous improvement. Our goal will be to use the analysis of metrics to achieve the following high-level objectives:

- Increase the amount of resource and staff working on process areas early in the IT operations lifecycle, where defect repair costs are the lowest. In other words, move key staff into pre-production engineering roles.

- Increase the amount of time spent on proactive and planned work, instead of reactive and unplanned work.

- Increase organization productivity by increasing change rates, change success rates, and the business value of changes.

- Keep closing the loop and using detective controls to carefully reduce variance including configuration variance, variance between planned work and actual work, and variance between builds.

Note that we focus not on service levels, but on the qualitative nature of the work being done. This is because service levels are a symptom of the quality and efficiency of an IT organization. Below, we list sample metrics for each of the three key process areas, followed by an example of some of the general improvement projects that you can take on to achieve the goals stated above.

Release Metrics

- **Time to provision known good builds**—How long does it take to build and provision infrastructure from bare-metal? (Shorter is better, and should be shorter than any MTTR requirement.)

- **Number of turns to a known good build**—How many times must the build be modified before it is acceptable for deployment? (Lower is better. A high number indicates the need for a more automated process.)

- **Shelf life of builds**—How long will each build be in production until it is replaced? (Longer is typically better, because it enables release management teams to stay out of reactive mode.)

- **Percent of systems that match known good builds**—According to the detective controls, how many production systems actually match their corresponding golden builds? (Higher is better, because it indicates absence of uncontrolled production configuration drift.)

- **Percent of builds that have security sign-off**—How many configurations were approved by security? (Higher is better, because it indicates that security is involved in the standard "blessing" process.)

- **Number of fast-tracked builds**—How many builds were rushed into production via the emergency change process? (Lower is better, because each of these represent a deviation from intended process.)

- **Ratio of release engineers to system administrators**—What percentage of staff is deployed on pre-production processes? (Higher is typically better because the cost of defect repair is much lower in pre-production.)

Controls Metrics

- **Number of changes authorized per week**—How many changes, as measured by the change management process? (In general, higher is better, as long as the change success rate remains high as well.)

- **Number of actual changes made per week**—How many changes, as measured by detective controls? (In general, higher is better, but should not be higher than the changes authorized by the CAB!)

- **Number of unauthorized changes**—How many changes circumvented the change process? This is typically measured by using the detective controls, or worse, through unplanned outages. (Lower is better.)

- **Change success rate**—How many changes are successfully implemented, without causing an outage or episode of unplanned work. (Higher is better: Best in class does better than 99%.)

- **Number of service-affecting outages**—How many changes result in service impairment or an outage? (Lower is better.)

- **Number of emergency changes**—How many changes required using the CAB/EC process. (Lower is typically better, since it indicates a higher percentage of planned work.)

- **Number of "special" changes**—How many changes, for whatever reason, are being made outside of the change process? (Lower is better, because these indicate that a change process is not fully functional, because management is allowing certain categories of changes to bypass change management entirely.)

- **Number of "business as usual" changes**—How many low-impact changes were there?

- **Change management overhead**—How much effort (in hours or staffing) is the change management function consuming? (In general, this number should be low. A high number may indicate a bureaucratic process, rather than one that enables productivity.)

- **Changes submitted vs. changes reviewed**—What is the ratio of evaluated change requests against the total change requests turned in?

Resolution Metrics

- **MTTR**—Mean Time To Repair: the average time to restore service after an interruption.
- **MTBF**—Mean Time Between Failure: the average time between service incidents.

Other Improvement Points

Release Area Improvement Points

- Enforce a standard build across all similar devices.
- Track all configurations in use (Development, Test, and/or Production) and ensure that there are stored builds for each.
- Do bare-metal builds whenever possible.
- Perform change audits on all production systems. Use detective controls to assure that all builds in production tie to known good builds. When they do not, investigate and mitigate.
- Segregate the development, test and production systems. Developers need systems they can use without fear of disrupting other activities. Test systems must reflect production configurations and be able to be locked down for controlled tests. Production systems, of course, must remain thoroughly controlled in terms of access and the change processes followed.
- Capture the known good state or "golden build" as part of the release management process.
- Create a library of automated build systems for all critical devices (i.e. repeatable, automated processes that can provision/re-provision all critical systems).
- Create a definitive software library (DSL).
- Define a process for accepting applications into, and retiring them out of, the definitive software library (DSL).

- Confirm that all deployed system images are under version management.
- Create regular meetings to determine the relevance of the repeatable build inventory.
- Confirm that repeatable builds have predictable and bounded remediation times.

Controls Improvement Points

- Change management meetings must have a specified agenda.
- Changes should be categorized (e.g. business as usual, application change), with appropriate workflows.
- Work with audit to agree on what reporting information should be provided, and learn what they will audit and inspect.
- If the rate of change is increasing beyond control, investigate the causal factors.
- If the rate of change is decreasing, confirm that staff is not circumventing the change process.
- Automate key change management processes.
- Create a feedback loop from production to the release engineers. Often, engineering groups get out of sync with the "real world" in terms of what works and what doesn't.
- Create and use your CMDB to track production infrastructure—make sure it is used by as many staff as possible as a central knowledge base.

Resolution Improvement Points

- Internalize the fact that change, regardless of source, is the root cause of most remediation efforts.
- Internalize the fundamental relationship between MTTR and availability. By improving MTTR, you also improve overall availability.
- Constantly improve problem diagnosis processes, since this is the longest part of the repair cycle.
- Define bulletproof rollback processes to recover from failed or unauthorized changes.
- Make sure that the problem ticketing system can, for affected systems, show all open work orders and production changes, both authorized and unauthorized.
- During remediation, use the CMDB and change management systems to see what has caused failures in the past.
- Track who made a specific change, enforcing accountability as well as fostering knowledge transfer.
- Have a process that maps all detected changes to a valid business purpose or authorized work order. Have the Change Advisory Board review this information as part of their regular meeting.
- Identify a set of change owners for all critical systems.
- Track repeat offenders who circumvent change management policies. Determine the best course of corrective action, starting with additional training and escalating up to and including formal disciplinary action.
- Review problem tickets during the change management meetings in order to identify required actions.
- All repairs made in the production environment must be mirrored in the preproduction environment, CMDB and the repeatable build process.

- Give systems and services meaningful names that relate to the functionality they provide.

- Create an end of shift audit process where operations managers are held accountable for changes that happened on their watch: Changes should either be mapped to authorized work or rolled back.

A Caution About Automation

When it comes to automation, IT process consultants usually warn organizations not to automate a new process before you have had practice running it with pencil and paper. For a historic example to put this in perspective, we can look to the challenge of bringing quality to automotive manufacturing in the 1980s. In the book Why Smart Executives Fail, Sydney Finkelstein describes how GM attempted to solve their quality issues through robotic automation. By the end of GM CEO Roger Smith's program, they had spent $44 billion to build the "factory of the future"—enough to have purchased Toyota and Nissan combined, but without meeting their quality or cost goals. Ford President Phil Benton was not surprised and reflected that consistency of practice must come before automation.

When dealing with any IT operational processes, whether it is related to change, configuration, release or security, take heed. If you cannot run these processes manually, do not attempt to automate it—all you will do is automate confusion.

What You Have Built And What You Will Likely Hear

> "Phases Three and Four of Visible Ops focus on creating a repeatable process and striving for continuous improvement. Visible Ops bundles all of the service life cycle in these two phases, and there is obviously a lot more to it than you probably intend to cover. Visible Ops is imbedding the four aspects of a service life cycle [Planning/Definition/Implementation/Supporting] in phases Three and Four. These phases merit attention in mature organizations that are beyond the phases of critical care. This is then also the beauty of Visible Ops: it points out to organizations where the start point is, and how to get out of the chaos."—JAN VROMANT, ITSM CONSULTANT

At this point you have created a closed-loop management system that uses metrics and controls to improve over time and better react to the environment. Points to consider:

- Auditors hate pushing organizations to implement controls, especially if doing so creates grudges and literal interpretations of findings. Ideal controls are owned by the business to meet objectives instead of being there only to generate positive audit findings.

- The closed-loop management system constantly reinforces the culture of causality.

- Objective data backs up assertions of improvement and issues that need to be dealt with.

- As opposed to management by belief, you have firmly moved to management by fact.

Summary

Congratulations! If you've made it this far, you've learned all about the four phases of Visible Ops. We hope you've found many practices that you can quickly implement in your own organization. Moreover, we know that many of you will not only implement these practices, but also improve on them—if you are one of these, please make sure to email us and tell us about it!

Based on our numerous years of research, we are confident that if you follow the steps outlined in this book, you will be able to replicate the amazing transformation that other IT practitioners have achieved with their organizations to fix their availability and security issues. We not only hope that you have found this journey to be useful, but that it has illuminated a different way to approach solving IT operational process issues.

IT practitioners can use Visible Ops to evolve from its artisan roots to a system of repeatable and verifiable processes, where security, availability, quality and value are built into the processes, instead of being inspected only at the end. If this sounds familiar, there's a good reason. These are the themes that Deming espoused throughout his career. As IT practitioners adopt a process-focused approach, more of Deming's genius will seem applicable on a daily basis. Here are some of his quotable gems:

- "If you can't describe what you are doing as a process, you don't know what you're doing."
- "It is not enough to do your best; you must know what to do, and then do your best."
- "Does experience help? No! Not if we are doing the wrong things."
- "We should work on the process, not the outcome of the processes."
- "Learning is not compulsory...neither is survival."

High-performing IT organizations clearly have a passion for process and consistency. They integrate security and IT operations just as the automotive industry successfully integrated quality into manufacturing. These organizations have created the culture necessary to ensure consistent, repeatable and verifiable IT operational processes, where both the security and operations groups are motivated to detect and reduce operational variance. These practices bridge the gap between security and IT operations, and also ensure that they always work together to achieve common objectives and business goals.

It is our sincere hope that the Visible Ops methodology will help your organization in your process improvement journey.

Sincerely,

Kevin Behr
Gene Kim
George Spafford

Information Technology Process Institute (ITPI)

This handbook was developed by the Information Technology Process Institute (ITPI). The ITPI is a not for profit organization, engaged in three principle areas of activity: research, benchmarking and the development of prescriptive guidance for practitioners and business executives. The ITPI has collaboration agreements in place with research organizations such as the Software Engineering Institute at Carnegie Mellon University and the Decision Sciences program at the University of Oregon. We are currently developing the necessary guidance that solves the common objectives of IT Security, Corporate Governance, Audit and Operations. Through research, development and benchmarking, the ITPI creates powerful measurement tools, prescriptive adoption methods, and control metrics to facilitate management by fact.

For more information please contact the ITPI at:

ITPI
2896 Crescent Avenue
Eugene, Oregon 97408
Main Telephone: (541) 485-4051
Main Fax: (541) 485-8163
http://www.itpi.org
info@itpi.org

The ITPI also runs the Community of Practice Listserv (ICOPL). The intent of this email list is to create a forum where we can exchange ideas, solutions, works in progress, and advance the cause of how IT operations, security, audit, and management can work together to solve common objectives. If you are interested:

• To subscribe: send a blank (subject and body are ignored) e-mail to icopl-subscribe@itpi.org.

• To unsubscribe: send a blank e-mail to icopl-unsubscribe@itpi.org.

• To email the list, send email to icopl@itpi.org.

• This page is reproduced at http://www.itpi.org/home/icopl.php.

Appendix A: **Preparing For Audit**

For many IT practitioners, the entire audit function may be one of the most mysterious and misunderstood roles encountered when working in certain industries. All too often, auditing is viewed as a necessary evil, and therefore, should be characterized by a confrontational relationship. Yet in high-performing organizations, there is a mutual respect between the IT teams and the internal and external IT audit teams they work with. In these situations, IT teams view auditors as additional resources to ensure that appropriate controls are in place and effective.

Typically, painful audit findings reveal the absence of effective processes and controls rather than being the fault of the auditors. Just as the manufacturing world realized the need for quality control processes, IT is finally recognizing that processes and controls must be implemented as well. When these processes are well-documented, and documentation exists that can demonstrate the controls are working, audits usually go much more smoothly, because auditors have a readily identifiable desired state to audit against.

When processes are not documented, auditors must grade you against their own processes. Worse, when controls do not exist to demonstrate that the processes are being followed, auditors must go into "archaeology" mode to determine for themselves if the systems meet documented control objectives. For example, if you claim to have change management meetings but do not have any formally recorded meeting notes, how can auditors verify that the meetings actually happened?

The level of documentation must be commensurate with the risks associated with the changes. A modification to a report heading is likely to have minimal risk associated with it. A substantial re-write of the Enterprise Resource Planning (ERP) system is a very different matter, has much more risk and thus requires additional processes and proof that those processes are effective. Resist the urge to document everything and instead focus efforts on creating evidence that the right processes are in place and are being followed.

To explain this further, we will examine the way auditors view the world, which is through three broad categories of controls.

Controls 101

Auditors often view the world through the lens of risks and controls. Risks exist, and you can mitigate them by either preventing or detecting them, and you should always be able to make corrections and recover should the risks actually happen. To explain this better, here are the three categories:

- **Preventive**—controls that keep something from happening. For example, policy, separation of duty, and authorization processes are all preventive controls.

- **Detective**—analytical controls that monitor activity and processes to determine if the preventive controls have failed or if something is out of compliance. For example, change monitoring and verification are detective controls.

- **Corrective**—corrective controls restore the situation back to the expected state. For example, if a system crashes due to a failed change, reloading all applications from the last known good image to bring the system back online serves as a corrective control.

The combination of the three types of controls creates a system of checks and balances to help ensure that the processes, people, and technology operate within prescribed bounds. We provide two simple examples of controls to reduce the risk of financial fraud and uncontrolled IT changes.

Business Risk	Preventive Control	Detective Control
Financial fraud within vendor payment process by someone creating a false vendor, and then paying themselves with fake purchase orders.	Separation of duty: those who can create vendor accounts cannot also issue payments to vendors. Authorization: vendor payment requires authorization from budget owner.	Review payment authorization for approved signatures.
Uncontrolled or unauthorized changes being made in the production IT environment thus jeopardizing availability, integrity and security.	Separation of duty: pre-production staff cannot access production systems, and must submit proposed changes via the change management process. Authorization: all changes are reviewed and authorized by the CAB.	Monitor production configurations for changes to guarantee that all changes map to an authorized work order.

Separation of duties ensures that no single person has complete unchecked access to do unauthorized things. Because lack of segregation can create nearly endless opportunities to commit fraud, developers are not allowed access to production processes where they can directly make changes in regulated environments. Instead, they must develop the code, then forward it to testing. Once there, the operations team can review the change, assess risks and deploy it into production if everything is acceptable.

These days, many audit concerns are driven by regulatory compliance needs that are required by the industry or that assure the integrity of financial reporting. At one time, auditors and bean counters checked to see if the financial statements were correct by opening up all of the warehouses and counting all the "beans." In this way, they could verify that the financial statements matched what they physically observed.

However, even the best auditors have finite time and resources. Instead of going into the warehouse and counting beans, they go to the bean counting machine and check the controls to determine whether the machine can be trusted. In most cases it is best to have a combination of preventive and detective controls. If neither exists, or if they are inadequate, the auditors cannot trust the results of the bean counting machine. This is a very bad thing, because it erodes their ability to rely on anything the bean counting machine did, and requires opening up the warehouses, and, guess what, counting the beans. In other words, without assurance that proper controls exist, far more scrutiny is required thus incurring substantial costs.

To create a more productive working relationship with auditors, be able to clearly describe your preventive processes and the detective controls that prove they work as expected.

A main premise of this book is that controls serve an important purpose to ensure that our processes achieve the desired business objectives and that controls are not in place simply to make generate positive audit findings or to comply with regulations. After all, a customer would not feel safe if the restaurant only complied with health codes just to keep the inspectors at bay. They would be happier if the restaurant handled food with care to keep customers healthy, happy and improve their overall dining experience.

IT is no different. An organization that uses effective controls to improve their processes typically has far better availability, lower amounts of unplanned work, better security, and incidentally, smoother audits.

Auditors—Internal And External

During your journey, you will encounter both internal and external auditors. Neither one are considered part of the management structure. In other words, management makes decisions, manages risks, and runs the business. Audit ensures that risks are managed and statements that management makes are reliable.

According to professional standards promulgated by the Institute of Internal Auditors, the internal audit function reports directly to the Audit Committee of the Board of Directors. Organizationally, internal audit staff may be placed under the CEO or CFO, but they are independent of business management. Typically, internal audit reports are addressed to management and copied to the Audit Committee.

External auditors, on the other hand, are third parties retained by management to give an unbiased opinion of the assertions made by management and report to the audit board and board of directors. Whereas internal audit reports to the audit board, external auditors are accountable to shareholders, regulators, and potential investors. External auditors come in and evaluate the effectiveness of the controls that are attested to being in place by the company's senior management. If a weakness is found, they include it in an audit findings report. If they find a material weakness, the auditors may be required to disclose their findings to the appropriate regulatory body, such as the Public Company Accounting Oversight Board (PCAOB).[14]

The Sarbanes-Oxley Act Of 2002

In the United States, the Sarbanes-Oxley Act, which aims to restore public confidence in financial reporting, has generated a great deal of activity. The law is broken down into sections and section 404 tends to generate the most concern in IT circles.

Section 404 is titled "Management Assessment of Internal Controls." The section states that it is, "...the responsibility of management for establishing and maintaining an adequate internal control structure and procedures for financial reporting." Since IT is so very pervasive in organizations today, especially finance, there must be appropriate controls. As mentioned earlier, it is unrealistic for auditors to inspect everything. They must rely on the presence of effective controls for security, availability and data integrity. Furthermore, external auditors must attest to the controls. A weakness in internal controls could trigger a disclosure to the SEC that there is a control deficiency. As investors would most likely view these disclosures negatively, a public company could see a negative impact on its stock price and the possibility of a shareholder lawsuit. Needless to say, this makes many companies very nervous.

For groups using the ISACA's Control Objectives for Information and Related Technologies (COBIT), please recognize that Visible Ops provides guidance on how to implement some of the controls set forth in COBIT. COBIT is very good at identifying what needs to be done, but implementers often require assistance from best practice sources such as Visible Ops and ITIL. You will find that Visible Ops provides guidance relating to controls identified in AI6—Manage Changes, DS9—Manage the Configuration, and M1 Monitor the Process.

---

[14] Please note that materiality does not always apply. If an opportunity for fraud exists and there is no control, then that is very serious because fraud signifies a control weakness. Whether $10 is taken or $10,000, a compensating control must be in put place.

Increasing Your "Auditability"

Auditors do not like to see organizations that simply have controls in place to pass audits. They much prefer to see organizations who embrace controls to improve the business. To this end, they look for documentation of processes and proof that controls are followed. The first three phases of Visible Ops covered these points:

1. Ask the auditors what they are looking for before an audit. Ask them for their audit objectives, if any pre-audit checklists or data will be required beforehand, what meetings are required, specific areas they will inspect, and so on. Most likely, they will explain what they will be looking for, and give you an opportunity to find out which processes and policies you will need to supply documentation. Remember, it is better for you and easier for the auditors if you can articulate against which target you wish to be measured.

2. Again, asking questions in advance is one of your best means for preparing for an audit.

3. It is better to be prepared for an audit and not need material than to have an audit and wish you had material.

4. Make sure to list your perceived risks. Back it up with a list of risks sorted in descending order with the highest risks at the top, along with the controls you created to mitigate them.

5. Document your preventive controls, and have detective controls in place to show they work.

   a. Document the change management process.

   b. Use the CAB meeting minutes to show that meetings are being attended and used to manage change.

   c. For each authorized change, document the configuration changes from the detective controls to show that the changes made were within the scope of the work order.

   d. File the data collected about change requests and make it readily accessible. In some organizations, all of the above information lives in a physical three-ring binder.

6. Keep a current and accurate asset inventory of hardware and software.

7. Document all internal audit procedures and the proof that they are being followed. For example, if your policies state that firewall logs are monitored by a system with exceptions reviewed, then you must have proof of following that policy through logs of one form or another.

8. Document all outages and unscheduled downtime in the systems along with corrective actions taken.

9. Keep current documentation of all exceptions to policies.

10. List any security incidents along with corrective actions taken.

11. Be able to produce previous audit findings, analysis of the findings and progress made against findings that warranted corrective action.

You will immediately notice that it is virtually impossible to prepare for an audit at the last moment. This is why you must develop a process culture that naturally produces the above-listed elements. For example, the documentation from change management meetings helps track what is going on and capture knowledge. It is not simply there to pass audits. When we discuss preparing for an audit, we are highlighting the minimum that auditors will want to see in the form of electronic or manually captured data. Again, the internal and external audit groups can best tell you what they need to perform audits.

The following summarizes our audit tips from each of the first three phases:

Phase I: Stabilize The Patient and Modify First Response

- Avoid at all costs creating an adversarial relationship with auditors. Instead, demonstrate that you have effective management and control processes in place, and the documentation to prove it. If you cannot show intended and actual activities, auditors go into "archaeology" mode. (The worst thing you can do is become defensive and adversarial, especially if material control weaknesses do indeed exist.)

- Make sure you have an up-to-date document describing your change management process. Show this to auditors up front to illustrate what you want to be measured against. Without it, they will bring in their own processes to measure you.

- Take good meeting minutes during the CAB meetings and file them. Make sure they are dated. Showing meeting minutes to auditors to demonstrate that the meetings are actually taking place.

- The mantra of post-Enron auditors is, "If it's not documented, it doesn't exist." Therefore, be sure to document both your work and your meetings. The correct level of documentation should be commensurate with the level of risk associated.

- To show that your change management processes function, meeting minutes should show:
  - Newly authorized and scheduled change requests.
  - Acceptance of implemented changes, showing correlation between detected changes and implemented changes, showing successful implementation, acceptance by a change manager and closure of the work order.
  - Changes to production equipment tracked in work logs/work order tickets. These should identify the date, time, implementer and system along with details of the changes made.

- Assemble a list of changes made outside of the change management policy and corrective actions taken.

- On a regular basis, create and review a report with the number of changes requested, changes approved, MTTR and Change Success Rate by asset, functional area and organization, etc.

- Engineer the change workflow and ticketing systems in such a way that "closing" a request or ticket is not possible until it has been reviewed and accepted by the change manager. This ensures accountability, visibility and fact-based management, instead of belief-based or faith-based management.

- By doing the above, you prove that you have functioning preventive, detective and corrective controls in place. For more information, refer to Appendix A.

Phase 2: "Catch & Release" and Find Fragile Artifacts Projects

- Be able to show the process of how you generated the catalog of services and the assets that support them. Remember, the inventory should include both hardware and software.

- Show that you understand the business processes you are supporting by working with senior management to rank the services by importance to the organization and their degree of fragility.

- Show how you assure that the inventory is maintained and accurate. The lack of an accurate inventory may indicate to auditors that there are inadequate controls.

- Show the list of fragile artifacts resulting from this phase as evidence that you are performing risk mitigation.

- Be able to document the systems and processes used to detect changes.

Phase 3: Create A Repeatable Build Library

- Fully document the build process from feature request, to build definition, to build acceptance.

- Fully document the acceptance and handoff process between the pre-production and production teams.

- Prepare reports on production rollouts of software, change success rate, time required to complete the rollout, and the integration with the change management processes.

- Document the process of how software is evaluated, accepted into, and purged out of the DSL.

- Generate a report of the percentage of deployed systems that match the golden builds.

- Document the process used to track threats and generate projects in the release management processes for patch updates and software rollouts.

- Document the policies for the clean-room build process.

- Be able to show how systems are certified. In other words, "How do I know that what I built is what I intended to build?"

- Be able to provide a list of all exceptions to the golden builds and justifications for them. An abundance of unexplained exceptions is evidence of an ineffective process.

Auditor Red Flags And Indicators

Clearly, auditors are concerned about the health of the IT systems hosting applications that the organization relies on. Thus, they focus on poor service levels and unusually high velocity of change as "red flags" indicating that there are inadequate controls. This is the profound concept that came from our study of high-performing organizations. The controls that give you good service levels are exactly the same controls that auditors look at to mitigate risks.

Appendix B: **The Information Technology Infrastructure Library (ITIL)**[15]

Many executives express frustration as they attempt to reign in the chaos and expense associated with their IT investments but find little in the way of substantive guidance. The IT Infrastructure Library (ITIL) has emerged as the worlds most widely accepted approach to the management and delivery of IT Services.

If you have not heard of ITIL, do not be surprised. ITIL currently has over 42,000 certified consultants, primarily in Europe and Canada, with only a small fraction of those certified professionals residing or practicing in the U.S.

The Information Technology Infrastructure Library (ITIL) represents a drastically different approach to IT by framing all activity under two broad domains named "Service Management" and "Service Delivery" respectively. By focusing on the critical business processes and disciplines needed to deliver services around IT, the ITIL provides a maturity path for IT that is not based on technology. This accessibility allows senior executives to both sponsor and shepherd IT quality improvement efforts. The ITIL has become the most widely accepted approach to IT service management.

ITIL provides a comprehensive, consistent volume of best practices drawn from the collective experience and wisdom of thousands of IT practitioners around the world. By defining IT quality as the level of alignment between the actual services delivered and the actual needs of the business the library serves as a common point of engagement for IT and the other business units.

What ITIL Covers

To codify and organize the guidance contained in the ITIL, the British Standards Institute published the BS15000 as a code of IT Practice for IT Service Management. The BS15000 organizes all of the guidance from the ITIL into five distinct categories. The areas include: Service Design and Management Processes, Supplier Processes, Resolution Processes, Control Processes, and Release Processes.

---

[15] Adopted from: Behr, Kevin and Kim, Gene. "Why You Need To Know About ITIL®." The original article was previously published and is available at http://www.bettermanagement.com/library/library.aspx?libraryid=5711&pagenumber=1.

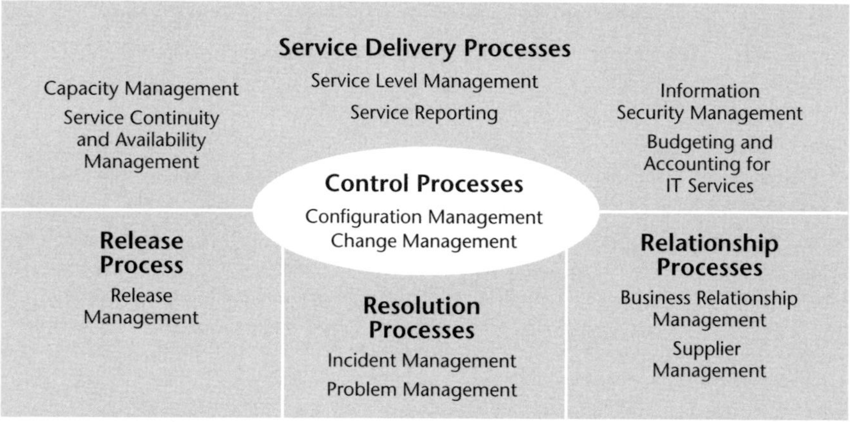

Service Delivery Processes

Capacity Management

Service Continuity and Availability Management

Service Level Management

Service Reporting

Information Security Management

Budgeting and Accounting for IT Services

Control Processes

Configuration Management

Change Management

Release Process

Release Management

Resolution Processes

Incident Management

Problem Management

Relationship Processes

Business Relationship Management

Supplier Management

Figure 7: The BS 15000 Diagram

- **Release Process**—This process area answers the question "Where does infrastructure come from before it is deployed?" This includes activities such as the planning, designing, building, and configuring of hardware and software. Unfortunately, release processes are traditionally the last process area that organizations invest in. Yet this is the process area that delivers the highest return on investment, because it encompasses the entire pre-production infrastructure, where the cost of defect repair is lowest.

- **Control Processes**—This process area covers maintaining production infrastructure, not only to prevent service interruptions, but also to efficiently deliver IT service. This is done through change management, as well as asset and configuration management. BS 15000 defines change management as well as asset and configuration management as primary controls. As Stephen Katz, former CISO of Citibank, once said, "Controls don't slow the business down; like brakes on a car, controls allow you to go faster."

- **Resolution Processes**—This process area is triggered when production infrastructure does go down, service is interrupted, or there is a security issue. Incident management owns the customer relationship, and problem management owns the tasks of turning each problem into a known error that can be more efficiently resolved the next time it happens. All too often, organizations that spend too much time firefighting are unable to spend time in the previous two process areas.

- **Relationship Processes**—This area focuses on the processes necessary to support effective customer relations as well as the management of third party vendors from a performance and contractual standpoint.

- **Service Delivery Processes**—The goal of these processes is to provide the best possible service levels to meet the business needs of the organization. This process area includes the monitoring and management of IT infrastructure as it relates to Security Management, Availability and Contingency Management, Capacity Management, Financial Management and Service Level Management and Reporting.

ITIL Success Stories

Now that you have been exposed to the ITIL and BS15000, the next logical question is "has anyone successfully adopted this framework?" The answer is a resounding yes. Over the last ten years organizations ranging both in size and industry have successfully integrated the best practices guidance and taken their results to the public. Some of the most familiar names include Procter and Gamble, Guinness, IBM, British Airways and the Internal Revenue Service. Procter and Gamble credit their adoption of ITIL practices with saving the company well over one hundred million U.S. dollars each year. Other organizations, such as Shell, have leveraged the ITIL and saved both large amounts of man-hours and hard currency.

It has become clear that focusing on the development of a process-driven IT organization can yield significant efficiency related savings. Effectiveness of services delivered has also ranked highly as one of the many positive outcomes of best practice adoption. Moving into the realm of IT Service Management requires the attention and focus of the IT organization as well as from many important stakeholders in other business units.

Easy Steps to Get Value with ITIL

How do you get started with ITIL? After all, ITIL includes volumes upon volumes of manuals that sometimes read like a dictionary. A better approach would be to look at several key process areas in your organization, to assess where you are versus best practices, and to generate smaller projects, bootstrapping controls and processes that need attention.

Below, we include a short questionnaire in several key areas that have plenty of leverage, and a strong probability of quick improvements with high visibility of return. Use the questionnaire to discover issues that commonly cause widespread service disruption and firefighting behaviors.

Control processes

1. Do your system administrators spend too much time fire-fighting?
2. Do you have a well documented change management process that provides visibility and control points for changes?
3. Are the largest percentages of problems caused by changes made internally as opposed to externally?
4. Can you quickly discover unauthorized or undocumented changes?
5. Can you lock down production servers so no change is allowed?
6. Can you map systems changes to a change request or an authorized work order?
7. Do you schedule all system changes to fixed maintenance windows, mitigating the risk of changes?
8. Do you have an end-of-shift audit process, assuring that the operations manager is handing over the data center in the same condition that it was received?

Release Processes

1. Can you enforce a standard configuration build across all of your devices?

2. Do you know precisely how many different configurations you have in your environment?

3. Can you reliably rebuild servers that are in production (i.e. "we can do bare-metal builds")?

4. Do you perform change audits for all of your production systems?

5. Are production and development systems clearly separated?

6. How do you ensure that the staging environment matches the pre-production environment before deploying builds into production?

7. Can you capture the known good state or "golden build" as part of the release management process?

8. Do you have confidence that the deployed systems match the golden build?

9. Do you have a library of automated build systems for all critical devices (i.e. repeatable, automated processes that can provision all critical systems)?

Resolution Processes

1. Is change regardless of source the root cause of most of your remediation efforts?

2. Is the longest part of your repair cycle spent diagnosing what is wrong?

3. Can you quickly detect what changed on systems during problem resolution?

4. Can you perform precision rollbacks when undesired change is detected?

5. During remediation, can you see all authorized work orders pertaining to a target system?

6. During remediation, can you see all previous work tickets to learn what has caused failures on your system in the past?

7. Do you track the change success rate?

8. Do you have a process that maps all changes to a valid business purpose or an authorized work order?

9. Have you identified a set of change owners for all critical systems?

10. Do you have a list of repeat offenders who circumvent change management policies?

From Denial, Acceptance and Problem Solving

If you are like most IT operation organizations, you may have uncovered some issues with this short questionnaire. Do not worry, because these are common problems that ITIL was designed to solve.

Appendix C: Focusing Efforts With An Integrity Management Capability Assessment (IMCA)

The IMCA exam leverages lessons learned from ITIL, BS 15000 and high performing organizations to measure an organization in four key ITIL process areas: Release Processes, Control Processes, Resolution Processes and the Security Management portion of the Service Design & Management Processes domain. The exam is conducted during a one-hour interview session wherein numeric scores are assigned. The end result is a detailed report approximately 20 pages long that identifies strengths and weaknesses in the areas mentioned. The questions asked in Appendix B are indicative of the questions asked for the exam.

The principal benefits of an IMCA exam are that it helps to bring unspoken weaknesses to light and creates executive level champions. Without an IMCA, organizations can embark without the initial knowledge of process status, but this is akin to taking a shotgun approach to potential root causes. The IMCA can create a capability assessment that can then be analyzed to accelerate a return on your efforts.

More information is available at http://www.itpi.org

Appendix D: A Glossary Of Terms

The following glossary is an attempt at identifying possibly ambiguous terms used in this book. A current and more complete glossary is available on the ITPI Web site http://www.itpi.org.

- **Availability:** Typically stated as the ratio of the time that the system is operating within acceptable bounds divided by the total possible time.

- **Business As Usual (BAU) Changes:** These are changes that are regularly made during the course of business with readily known risks and outcomes are readily known.

- **Change Advisory Board (CAB):** A defined group of stakeholders with vested interests in the system in question who are able to weigh the risks and benefits of change while maintaining proper communication.

- **Change Advisory Board/Emergency Committee (CAB/EC):** When there is an urgent or emergency change and the entire CAB cannot be convened, this is a defined smaller group of stakeholders who can review the change request and make a proper decision as to implementation. Their decision should then be reviewed when the CAB next convenes.

- **Change Success Rate (CSR):** The ratio of successful changes to total changes. This metric shows the relative effectiveness of the change management process.

- **Configuration Drift:** The tendency for configurations to change over time. For example, a server's configuration is most certainly known when it is first released. As time goes on, patches get applied, and there is human intervention. As these changes accumulate, undocumented changes may have occurred and the actual configuration has thus "drifted" away from the known configuration.

- **Configuration Item (CI):** One discrete build that is tracked. It may be a base component that can not be further divided or an assembly made up of other configuration items. CIs can be hardware, software, documentation or a combination thereof.

- **Configuration Management Database (CMDB):** A system used to track configuration items, requests for change, work orders, errors, relationships, etc. The definition is often nebulous as the exact implementation varies across organizations. Fundamentally, it is the core system(s) that tracks all activities including service levels.

- **Defense-in-Depth:** A security strategy of using many layers of defense as opposed to relying on fewer layers, perhaps even just a single layer. The thought process is that by using layers, each provides an additional level of security should the preceding barrier be breached.

- **Definitive Software Library (DSL):** A repository of authorized software that is secure and has version control. Software may only be added or removed from the library through formal processes.

- **Detective Control:** Processes or systems that review records to determine activity. In IT, a change monitoring and reporting system that reviews configurations against known standards on a periodic basis and reports observed changes is an example of a detective control. Likewise, a manual review of a log file looking for anomalies is an example of a detective control.

- **Diff:** To create/detect a delta or "difference" between two items. Take lists "ABC" and "CDE". The diff is "ABDE" as they are unique to each list while "C" is common in both lists.

- **Domain Name Server (DNS):** A server application used to map hostnames to Internet Protocol (IP) addresses.

- **Dynamic Host Configuration Protocol (DHCP):** A protocol that automatically communicates network configuration settings from a central host to distributed assets, which are configured to use the protocol to obtain their Internet addresses, domain name servers, gateways, etc.

- **Forward Schedule of Change (FSC):** A schedule that contains details of all changes and their proposed implementation dates. Items are added to it through the approved change control process.

- **Integrity Management Capability Assessment:** See Appendix C.

- **ITPI Community of Practice List (ICOPL):** An email list maintained by the ITPI organization to facilitate knowledge transfer.

- **Information Technology Infrastructure Library (ITIL):** A collection of best practices codified in seven books by the Office of Government Commerce in the UK. http://www.ogc.gov.uk/index.asp?id=2261

- **Information Technology Process Institute (ITPI):** A not-for-profit organization engaged in three principle areas of activity: Research, Benchmarking and the Development of prescriptive guidance for practitioners and business executives.

- **Mean Time Between Failures (MTBF):** The average time between failures of the asset.

- **Mean Time To Repair (MTTR):** The average time it takes to restore service once an asset has failed or dropped below acceptable service levels.

- **Request For Change (RFC):** A formally submitted document, or electronic record, that identifies the relevant information surrounding the desired change.

- **Revision Control System (RCS):** An application that tracks versions of files or potentially another form of data through the use of access and check-in/check-out controls. Depending on the system, a baseline with subsequent differentials may be tracked or each new version is stored and assigned a unique ID.

- **Shelf Life/Release Shelf Life (RSL):** Defines how long a build will remain viable before becoming obsolete.

Appendix E: CMDB Table Structures

Assembling a CMDB can be done in many ways. The following table represents key fields to be included in any CMDB at the CI master level:

The following flags can be used to track the current status of each CI entry:

Attribute	Description
CI ID	A unique name identifier by which the CI can be identified (Company Name-Location-CI Type)
CI ID Number	A unique number generated by the CMDB
CI Description	A description of the CI
CI Category	General category of the CI
Owner Responsible	Name of person responsible for this CI
Customer	Company name of customer
Date Acquired	Date the organization took ownership
Status	Is the component currently registered, accepted, under development, installed, withdrawn, etc.
Next Maintenance Window	Date of next scheduled maintenance (if applicable)
Make	Manufacturer
Model	Model Name
Model Number	Model Number
Part Number	Hardware part number
Serial Number	Hardware serial number
License Number	Software license number
Version Number	Software version number
Source Supplier	Who provided the component?
Relationship	Parent/Child; CI is connected to another CI; CI is resident in another CI; CI using another CI
Relationship Number	CI IDs are used to create the Relationship Number.
Location	The physical location of the device. Data center-rack location-rack unit ID
Ticket Numbers	The ticket number of incidents, problems, and change requests related to this CI

The following flags can be used to track the current status of each CI entry:

Status Flag	Description
Ordered	CI has been ordered from a vendor but has not yet arrived and therefore cannot be registered.
Registered	CI has been received and fully identified in the MDB database.
Accepted	CI has been accepted by the CI owner. This designation means that the CI process owner has verified that the CI meets the specifications that were called out. This is a verification that a Quality Assurance process has occurred. The process owner during registration is the "Owner Responsible."
Development	CI is in the development environment.
Testing	CI is in the testing environment.
Installed	CI is in the production environment.
Under Change	CI is in the process of being changed.
DSL	CI is part of the Definitive Software Library.
DHL	CI is part of the Definitive Hardware Library.
Archived	CI is archived and under the control of Storage Management.
Obsolete	CI is obsolete.
Missing	CI is missing
Stolen	CI was stolen.

Appendix F: **References**

Behr, K. and Kim, G. (2003). "Why You Need To Know About ITIL." *BetterManagement.com*, URL: http://www.bettermanagement.com/library/library.aspx?libraryid=5711&pagenumber=1, last visited March 7, 2004.

British Standards Institute (2002). *BS 15000–1:2002—IT Service Management: Part 1: Specification of Service Management*, September, p. 2.

Finkelstein, S. (2003). "Why Smart Executives Fail: And What You Can Learn From Their Mistakes," *Portfolio*, May.

Fordahl, M. (2003). "Under-construction satellite topples to floor in mishap," *The Associated Press*, September 10.

Mell, P. and Tracy, M.C. (2002). *Procedures for handling security patches*, Gaithersburg: National Institute of Standards and Technology, URL: http://csrc.nist.gov/publications/nistpubs/800-40/sp800-40.pdf, last visited March 7, 2004.

Miller, D. (1999). "Hardware High-Availability Programs in Action. (Product Information)," *ENT News*, June, URL: http://www.entmag.com/archives/article.asp?EditorialsID=6753, last visited March 7, 2004.

Nattermann, P.M. (2000). "Best Practice Does Not Equal Best Strategy," *McKinsey Quarterly*, Number 2, URL: http://www.bettermanagement.com/Library/Library.aspx?a=11&LibraryID=8387, last visited: March 7, 2004.

Office of Government Commerce (2002). "Best practice for service delivery," *Information Technology Infrastructure Library (ITIL)*, Norwich: The Stationery Office Limited.

Office of Government Commerce (2002). "Best practice for service support," *Information Technology Infrastructure Library (ITIL)*, Norwich: The Stationery Office Limited.

Scott, D. (2001). *NSM: Often the Weakest Link in Business Availability*, Editor: Gartner Group, URL: http://www4.gartner.com/DisplayDocument?id=334197&ref=g_search, last visited March 7, 2004.

Software List

Automated build systems

AIX
NIM, URL: http://publib-b.boulder.ibm.com/Redbooks.nsf/RedbookAbstracts/sg245524.html?Open, last visited March 7, 2004.

Solaris
Jumpstart, URL: http://wwws.sun.com/software/solaris/8/ds/ds-webstart/#4, last visited March 7, 2004.

Windows
InstallShield AdminStudio, URL: http://www.installshield.com/products/adminstudio/, last visited March 7, 2004.

Change monitoring and reporting

Tripwire for Servers, Tripwire for Network Devices, URL: http://www.tripwire.com, last visited March 7, 2004.

Ticketing systems

Best Practical *RT/RTIR* (Request Tracker, Request Tracker for Incident Response), URL: http://www.bestpractical.com/rtir/, last visited March 7, 2004.

HP *OpenView Service Desk*, URL: http://managementsoftware.hp.com/products/sdesk/index.html, last visited March 7, 2004.

Remedy *ARS* (Action Request System), URL: http://www.remedy.com/solutions/coretech/index.html, last visited March 7, 2004.

Appendix G: High-Performing IT Organizations: What You Need to Change to Become One [16]

IT is being challenged on many fronts, from cost containment, business alignment, compliance, competitive pressures in managing outsourced IT services, and security. Many experienced IT practitioners confronted by this potentially staggering array of challenges will point out that the solution to virtually all these issues is more repeatable IT processes and effective controls. However, merely understanding this does not necessarily equate to an effective plan to solve the problems, and may create more questions than answers. To simplify the problem, Dr. Eliyahu Goldratt, creator of the Theory of Constraints, articulates three simple questions that must have credible answers: What do I need to change, what should I change to, and how do I cause the change?

Finding answers to those three questions has been an area of passion for Gene since 1999. He has been researching high-performing IT operations and security organizations, attempting to understand what makes them so different than typical IT organizations, as well as studying how organizations have accomplished the transformations that take them from being merely average to best in class. Along this journey, Gene started working with other organizations that are also interested in these issues, such as SANS, the IT Process Institute (ITPI), the Institute of Internal Auditors (IIA), and most recently, the Software Engineering Institute (SEI) with Julia Allen. In particular, the collaboration between the ITPI and SEI has yielded some extremely promising results, both in characterizing high-and low-performing IT organizations, the key differences in their belief systems, and the necessary components to achieve an organizational transformation from low- to high-performer.

In this article, we will discuss two areas of research that we believe are foundational for answering the question of what IT organizations typically need to change and what they need to change it to. We will present a working definition of what characterizes a high-performing IT organization, and then discuss the key differences in the belief systems between them and more typical IT organizations in three areas of pain: patch management, proliferation of IT management scorecards, and managing outsourced IT services.

Lastly, to help answer the question of how to cause the change, we will describe the publicly available Visible Ops methodology, which captures how IT organizations have transformed into high-performers in a way that is can be accomplished in four steps, each which is a finite project and returns more value back than was invested. We will also describe the ITPI Community of Practice Listserv, and the upcoming VEESC benchmarking study. We conclude the article with a call to action and an active solicitation for feedback in participation in creating this community of practice for high-performing IT organizations.

Key Characteristics of High-Performing IT organizations

Since 1999, after studying the IT processes of hundreds of organizations, it started becoming clear to Gene that a handful of them stood out as somehow different from the others in some notable way. He started keeping a list of these organizations, at that time informally called "Gene's list of people with amazing kung fu." In 2000, Gene started working with Kevin Behr, CTO of one of these unusual organizations, and they started a more

[16] Adopted from: Allen, Julia and Kim, Gene. "High-Performing IT Organizations:
What You Need to Change to Become One." The original article was previously published and is available at
http://www.bettermanagement.com/Library/Library.aspx?a=13&LibraryID=9429. Reprinted with permission from BetterManagement.com

systematic analysis of what was common to these organizations, which they renamed the "best in class IT operations and security organizations." In 2003, Julia Allen from the SEI actively joined this effort, which resulted in a remarkable event in October 2003 at Carnegie Mellon University called the Best In Class Security and Operations Roundtable (BIC-SORT). Among the stated goals were to "begin to build an executive-level community of practice for IT (information technology) operations and security, with a common sense of purpose and a desire to influence other relevant and connected communities of practice; and to better capture and articulate the relevant bodies of knowledge that enable and accelerate IT operational and security process improvement." Since then, we have been actively processing and synthesizing the data we collected.

Based on our analysis, we have created the following working definition of high-performing IT organizations: They are effective and efficient and they succeed in applying resources to accomplish their stated business objectives with little to no wasted effort. These organizations have evolved a system of process improvement as a natural consequence of their business demands. They regularly implement formal, repeatable and secure operational processes.

Results of informal benchmarking indicate that in these best-in-class IT organizations, IT operations and security work together to create higher service levels (e.g. as measured by mean time to repair, mean time between failure); higher percentage of planned, scheduled work relative to unplanned work; unusually efficient cost structures (e.g. as measured by server to system administrator ratios); productive working relationships with management and peers; and smoother audits. Furthermore, they have more timely identification and resolution of security incidents, the earliest integration of information security requirements in the service delivery lifecycle, and the ability to quickly return to a reliable and trusted operational state. And perhaps most admirably, these organizations devote increasingly more time and resources to strategic issues, having mastered tactical concerns.

The high-performing organizations desire to detect production variances early so they can fix problems in a planned manner and where the repair costs are lowest and have the least impact. They value repeatable and verifiable processes and use controls to improve efficiency and effectiveness. And because these organizations use controls to improve their own operation, life is much easier for auditors who evaluate operational risk based on the presence of effective and verifiable preventive, detective, and corrective controls. In other words, the controls aren't there just because auditors asked for them, but because they are used to improve daily operations! As a result, high-performing organizations require considerably less effort to meet management and audit expectations.

To achieve these characteristics, several key performance metrics are essential to this level of performance: they have the highest change success rate (typically over 98%), highest effective rate of change (sometimes making over 1000+ successful changes per week), highest level of mastery of production infrastructure (achieved by low configuration counts and low configuration variance), and highest ratio of staff dedicated to pre-production activities (achieved by release management processes, pre-production testing, etc.).

Surprisingly, we found that all of the high-performing IT organizations had independently developed virtually the exact same processes to achieve these results. They shared similarities in three key process areas, which we will describe in the parlance of ITIL (IT Infrastructure Library): they had a "culture of causality" that ensured all production problems ruled out change as early as possible in the repair cycle (resolution processes), they had a "culture of change management" embedded in the way all work is done (control processes), and they moved as many production changes through a pre-production process that orchestrated changes with the production environment (release processes).

The high-performing organizations all implemented virtually the same procedures in these three ITIL process areas, which form the minimal closed-loop that generates metrics that allow continual process improvement. These procedures and processes are described in the Visible Ops methodology in detail, published by the ITPI (http://www.itpi.org/home/visibleops.php).

Belief System Differences Between High- And Low- Performing IT Organizations

Given the fact that high-performing IT organizations exist, what prevents low-performing organizations from becoming high-performers, given the promise of a better way? Understanding why this was so became one of the main areas of activity after the BIC-SORT event. Julia Allen, Kevin Behr, and Gene Kim from the ITPI and SEI have been synthesizing the captured list of key areas of pain and promise from the participating organizations during BIC-SORT. Our goal was to create a taxonomy of pains, find any cause-effect relationships and root causes, and understand what belief systems that preserved the status quo for the low-performers.

In the BIC-SORT, we captured almost one hundred specific areas of pain, such as the challenges of keeping up with security patches, the massive efforts required to do effective audits of business peers, and so forth. Of these, we chose to analyze three of the most acute of the listed pains: keeping up with patching, dealing with the proliferation of management scorecards, and management of outsourced IT services.

To analyze these problems, we used a technique pioneered by Dr. Eliyahu Goldratt called the Theory of Constraints Thinking Tools, specifically problem clouds and current reality trees (for more information, see http://www.thedecalogue.com/Tools/crt.htm). The goal was to understand the causal factors and beliefs that led to the high- and low-performing behaviors, and then finally, to find any commonalities among the three pain areas. What we found was illuminating.

Volume of Security Patches

An area of pain articulated by many of the participants at the BIC-SORT was the volume of urgent patches needing to be applied to infrastructure, resulting from the constant stream of new security vulnerabilities, and the need to find an effective solution to managing patches.

In the low-performers, this activity was characterized as ad hoc, chaotic, and urgent. Announcement of the availability of a patch to address a critical security vulnerability would lead to widespread chaos and disruption, often resulting in massive amounts of unplanned work at the expense of planned work. Worse, even successfully deploying the patch would often cause unintended consequences, such as servers becoming non-functional or even non-booting.

In contrast, the high-performers addressed patching as a predictable and planned activity, treating them as just another change. Announcement of critical patches would result in merely adding the patch to the release engineering candidate queue, where it could be evaluated, tested and integrated into an already scheduled deployment. The absence of urgency and a well-defined process for integrating changes leads to a much higher change success rate. Interestingly, virtually all of the high-performers apply patches much less frequently than the low-performers, perhaps by one or two orders of magnitude!

Proliferation of IT Management Scorecards

BIC-SORT attendees also listed the proliferation of IT management "scorecards" and other management and assessment instruments as another area of pain. We also threw into this category all the various industry compliance requirements, ranging from Sarbanes-Oxley Section 404, Gramm-Leach-Bliley, HIPAA, and so forth.

In the low-performers, this activity was characterized by having to look to external sources and authorities for the desired behaviors and measurements. The absence of a strong internal IT management framework and belief system might lead to adopting a "scorecard du jour," or worse, multiple external scorecards simultaneously that conflict with each other. This would lead to more work for the organization, and excess retrofitting to deal with the necessary process and organizational gymnastics. Worse, executive turnover might result in switching scorecards, which repeats the entire chaos cycle.

In contrast, the high-performers have their own clearly defined performance goals and desired characteristics. If the need to conform to an external scorecard or regulatory requirement materializes, they assign a small team to demonstrate traceability to it. Consequently, they have a lower cost of developing, sustaining and documenting controls, a better posture of audit and compliance, and have little need to look externally for authorities to tell them how they need to operate.

Managing Outsourced IT Services

The last area of pain we analyzed was the challenge of managing outsourced IT services. Any challenges with IT are inherently made more complex when these services are provided by an outside provider instead of an employee: corrective actions may have contractual implications, the scope of corrections may be constrained by the service level agreement, and so forth.

In the low performers, there is often a real desire to transfer the IT risk and responsibilities to someone else, especially if management perceives an absence of internal skills to meet the business objectives. However, when IT functions are outsourced, such functions rapidly become out of sight and out of mind, until the organization finds that it is unable to control and attest to the controls of the service provider. The organization then discovers that it may have inadvertently exacerbated the challenges by outsourcing, but unfortunately, "re-insourcing" the services may no longer be an option.

In contrast, the high-performers manage outsourced IT services just like any other business unit or project. They understand the unique positive and negative challenges of fulfilling IT projects or services by an external party. As a result, they tend to develop more bullet-proof service level agreements to proactively get better service and create avenues for future corrections from the service provider.

Common Root Causes For Preservation Of the Status Quo

After analyzing the three areas of pain, we started looking for common patterns and root causes that led to the preservation of the status quo in the low-performers, despite the clear promise of achieving the characteristics of the high-performers. We found five areas of root cause.

1. The absence of explicit articulation of current state and desired state hides or obscures the amount of pain

Often, management will conclude that the current state, along with all the companion pains, is tolerable. These organizations may articulate a litany of pains and frustrations, but in the absence of being able to quantify the pain, may decide without that it probably does not hurt enough yet to warrant any corrective action. This may be because of a sincere belief that the pain is not high enough yet, or it may be the following:

2. A culturally embedded belief that control is not possible

Often, management may not know that there is an alternative, believing that the control is not possible due to its nature (i.e. "IT operational and security issues are like the weather: there is nothing we can do about it, and that bad things happen to us, just like rain or hurricanes"), control is not possible due to business needs (i.e. "my business environment is too dynamic to accommodate bureaucratic processes or controls"), or maybe even deliberate abdication of responsibility.

3. Rewards/reinforcements for personal heroics vs. repeatable, predictable discipline

Often, there may be a culture or a hidden reward system that encourages heroics and a "cowboy culture." For instance, one person may work throughout the night for an entire weekend fighting a fire and get rewarded as the hero who saved the day. What is overlooked is that if one person can save the entire boat, one person can probably sink it, too. In these organizations, implementing effective processes and controls may be resisted or actively ejected, almost as an immune system would resist an unknown and foreign object.

4. Continued argument that IT operations and security are different than other business investments or projects

Often, there may be a view that IT is different than other business functions or projects, thus leading to need to determine the "business alignment of IT." Worse, IT may be operating as a silo, but here may be a separate security silo inside it! There is a common belief that ongoing security can exist outside the scope of IT operations. While security requirements certainly exist outside of the IT context, security controls must be embedded into IT processes so that they are jointly owned by both the IT and security organizations. When the two organizations do not have defined roles where they are collectively solving common business objectives, blame-games and finger-pointing for failures can cause a downward spiral.

5. A desire for a technical solution, which is easier to justify and implement than people and process improvements

Often, because of their backgrounds, IT management values automation and technology over repeatable processes and controls. In the absence of properly functioning processes and controls, the massive deployment of security technology solutions invariably results in the staggering capability to automatically perform devastating, irreversible IT operational changes in mere seconds, resulting in potentially monumental episodes of unplanned work and chaos for the entire organization. Combined with the previous root cause, this factor creates the kindling for an extremely fast and accelerated downward spiral.

The entirety of these findings is available in a report published by the SEI as follows: Allen, Julia; Behr, Kevin; Kim, Gene et al. *Best in Class Security and Operations Round Table Report* (CMU/SEI-2004-SR-002). Pittsburgh, PA: Software Engineering Institute, Carnegie Mellon University, March 2004. Copies of the report are available upon request.

Summary and Call To Action

In this article, we explored three critical questions in the context of solving the most common IT challenges: What do I need to change, what should I change to, and how do I cause the change?

By studying high-performing IT organizations, the areas that most often need changing in lower performing organizations are those with cultures that sustain a belief that control is not possible, that the absence of controls have tolerable costs, that success of the individual can outweigh the needs for success of the organization, and that somehow IT security and operations are independent of each other. By overcoming these incorrect beliefs, and by implementing repeatable processes in the ITIL process areas of release, controls and resolution as outlined in the Visible Ops methodology, organizations can not only achieve a belief transformation, but a performance transformation as well.

So, here is our call to action: Do you agree or disagree with our definitions of high- and low-performing IT organizations? Do you have more characteristics that should be added to our list of best-in-class attributes? If so, please let us know by emailing us at genek@tripwire.com or jha@sei.cmu.edu.

Also, if you are interested in any of this work, please join the ICOPL mailing list. Subscription information is at http://www.itpi.org/home/icopl.php.

About Julia Allen

Julia Allen is a senior member of the technical staff within the Networked Systems Survivability Program at the Software Engineering Institute (SEI), Carnegie Mellon University (CMU). The CERT® Coordination Center is also a part of this program. Allen is engaged in developing and transitioning enterprise security frameworks and executive outreach programs in information security, survivability, and resiliency. Previously, Allen served as acting Director of the SEI for 6 months and Deputy Director/Chief Operating Officer for 3 years. In addition to technical reports for CMU/SEI, she is the author of The CERT Guide to System and Network Security Practices (Addison-Wesley, June 2001).

About The Authors

Kevin Behr

Kevin Behr is the president and founder of the Information Technology Process Institute (ITPI), as well as the CTO of IP Services. Kevin's 15 years experience in IT operations, security and field engineering spans environments ranging from financial services, manufacturing and technology sectors, allowing him to identify common problem domains and develop powerful solutions for IT Operations that span industry and scale. Kevin is working on development of IT operations management curriculum and research grants in conjunction with researchers from the Decision Sciences and MBA Programs at the University Of Oregon Lundquist College Of Business. Kevin is currently working with Gene Kim and Julia Allen, a senior member of the technical staff within the Networked Systems Survivability Program at the Software Engineering Institute at Carnegie Mellon University on prescriptive adoption methods that integrate best practices in IT operations, security, and audit. Kevin holds the Certified Information Systems Auditor designation and is also ITIL certified.

Kevin is also a frequently invited speaker called on to address a broad range of technology and management framework topics by organizations such as the National Academy of Science, Hewlett-Packard, The SANS Institute, AFCOM, the Palmer Group, the Software Engineering Institute at Carnegie Mellon University, CERT, Tripwire, and BetterManagement.com.

Gene Kim

Gene Kim is the CTO and co-founder of Tripwire, Inc. In 1992, he co-authored Tripwire while at Purdue University with Dr. Gene Spafford. Although Gene is widely published on computer security, operating systems and networking in SANS, ACM and IEEE publications, he is continually fixated on the problems of process integrity issues in Operations and Security. He is currently actively working on a series of projects with the Software Engineering Institute and the Institute of Internal Auditors to capture how "best in class" organizations have Security, Operations, Audit, Management, and Governance working together to solve common objectives. Gene is certified on both IT management and audit processes, possessing both ITIL Foundations and CISA certifications. In 2004, he was named by InfoWorld as one of the "Four Up and Coming CTOs to Watch."

Gene holds an M.S. in computer science from University of Arizona and a B.S. in computer sciences from Purdue University. Gene co-chaired the April 2003 SANS technical workshop called *Auditable Security Controls That Work*, hailed by SANS as one of their top five gifts back to the community. In October, Gene co-chaired the *Best In Class Security And Operations Roundtable (BIC-SORT)* with the Software Engineering Institute at Carnegie Mellon University.

George Spafford

George Spafford is an IT process consultant interested in the intersection of human factors, security, and complexity in the world of information technology. Despite the productive collaboration with Gene Kim, George has no relationship to Dr. Gene Spafford of Purdue University, who co-authored Tripwire in 1992. George is a prolific author on a wide range of topics including project management, technology business, communication, and security. He is the Vice President of Publishing for the IT Process Institute (ITPI), a non-profit organization whose goal is to further IT process improvement.

George has held a number of positions in IT operations, development and management. He holds an MBA from Notre Dame, a BA in Materials and Logistics Management from Michigan State University and an honorary degree from Konan Daigaku of Japan. He's a member of the ISACA and ITPI.

Subscribe to the ITPI

Information Technology Process Institute

Form of subscription

(circle one)   Individual   Company   Consultant   Institution

(Qualifying institutions are academic institutions and not-for-profit organizations. Subscription in this category is subject to acceptance by ITPI.)

Personal Details

Title: _____

Last/Family Name: _____ First Name: _____ Middle Initial: _____

Business Name: _____

Address line 1: _____

Address line 2: _____
(if required)

City: _____ State/Province: _____ Zip/Postal Code: _____

Country: _____

Phone: _____ Fax: _____

E-mail: _____

Other information (used only by ITPI to improve subscription services)

How did you hear about the ITPI?

❏ Word of mouth   ❏ Employer   ❏ Internet search   ❏ Conference
❏ Journal   ❏ Web seminar   ❏ Round table

Current field of employment

❏ Financial   ❏ Banking   ❏ Insurance   ❏ Transport
❏ Retail and wholesale ❏ Government (National)   ❏ Government (State or local)   ❏ Public accounting
❏ Consulting   ❏ Education (student)   ❏ Education (instructor)   ❏ Public accounting
❏ Manufacturing   ❏ Mining/Construction/Petroleum   ❏ Utilities
❏ Other service industry ❏ Law   ❏ Health Care   ❏ Other

What is your level of knowledge of the Information Technology Infrastructure Library (ITIL)?			
None	Novice	Advanced	Expert

Current professional activity (circle one)	**CEO**	**CFO**	**CIO/IS Director**	**COO**
	IS Security Director	IS Audit Manager	IS Security Manager	IS Manager
	IS Support Director	IS Support Manager	Audit Director/ General auditor	External Audit Partner/Manager
	Internal Auditor	IS Security staff	IS consultant	IS Auditor
	IS vendor / supplier	IS Educator	IS Student	Other

Certifications obtained

CISA	CISM	CPA	CA
CIA	CBA	CCP	FCA
CFE	MA	FCPA	CFSA
CISSP	Other		

Would you like information about the Integrity Management Capability Assessment (IMCA)?

Yes	No

Subscription Fees

Individual	1 year ($125)	2 years ($240)	3 years ($330)
Corporation	1 year ($800)	2 years ($1520)	3 years ($2700)
Consultant	1 year ($1500)	2 years ($2700)	3 years ($3700)
Institution	1 year ($400)	2 years ($800)	3 years ($1100)

Do you want to purchase additional IMCA licenses?

• Appropriate subscription discount included
• Total prices shown

Individual	1-pack @ $425	2-pack @ $725	4-pack @ $1325	12-pack @ $3525
Company	1-pack @ $400	2-pack @ $640	4-pack @ $1120	12-pack @ $3320
Consultant	1-pack @ $400	2-pack @ $640	4-pack @ $1120	12-pack @ $2880
Institution	1-pack @ $500	2-pack @ $800	4-pack @ $1400	12-pack @ $3600

TOTAL $

Check or money order is currently the only payment option. Please make checks (US dollars only) out to ITPI. Send this form with payment to ITPI, 2896 Crescent Avenue, Suite 104, Eugene, OR, 97408

IMCA

Integrity Management Capabilities Assessment

What is it?

IMCA (Integrity Management Capabilities Assessment) is a powerful benchmarking tool, based on Information Technology Infrastructure Library (ITIL) best practices for IT and the BS 15000 Code of IT practice. The assessment was developed with two primary goals in mind. First, to capture a clear and detailed picture of an organization's current strengths, weaknesses and areas of risk with regards to the integrity management capabilities required to maintain a secure, stable IT environment. Second, to present specific recommendations based on an in-depth analysis of the unique operational challenges and IT environment being measured.

What is the process?

The assessment begins with an interview process that takes approximately one hour. The completed questionnaire goes on to produce an executive summary that scores your operation in several key ITIL process areas: Release, Controls, Resolution and Security. Your results are then compared to other companies in your industry as well as industry best practices, with the final results delivered in a comprehensive document (approximately 20+ pages). This evaluation is used to develop a recommended roadmap that provides specific process improvements. The improvements include the prescriptive adoption of specific best practices that are relevant to your operation.

What results can I expect?

Have you ever wanted to objectively measure the efficiency of your IT operations? Many of our clients have expressed concerns about their ability to quantitatively analyze the effectiveness of their IT operations. After conducting the IMCA, these organizations were able to identify common "best in class" characteristics where Security, Operations, Audit, and Management can work together to reach common objectives.

Organizations have benefited from IMCA through improved integration of Security into Operations, and have attained improved operational service levels and efficiencies, as measured by:

- Improved Server/SysAdmin ratios
- Lower Mean Time To Repair (MTTR)
- Higher Mean Time Between Failures (MTBF)
- Reduced aggregate downtime
- Decreased security risks with increased control
- Shortened provisioning times
- Maximum change and configuration management integrity

Who should be involved?

To ensure maximum value from the IMCA process, companies should plan to include at least two of the following people:

• Vice President of Operations (or other sponsoring executive)

• Line Operations Manager

• Change Control Manager

• Representatives from the Security staff

It is also recommended to assess a group that all report to the same business area/unit.

Areas Evaluated by the Assessment

Control Processes:

The assessment examines your organization's capabilities relating to operational controls, ranging from the ability to detect changes made to infrastructure to the evaluation of processes in place for change management. An effective change management process requires review of all proposed infrastructure and software changes before they are made. This review is typically performed by stakeholders such as Security, Operations, R&D as well as Internal Audit. The team decides whether the proposed changes are appropriate and include the necessary safeguards to ensure continuity of service.

The assessment also examines configuration management practices that guarantee all critical software and configurations stored match those running in production. This ensures a known, good repository for all standard configurations used across the enterprise, and guarantees that as configurations change, new configurations are documented and supersede the older revisions.

Resolution Management Processes:

This part of the assessment analyzes the company's problem management processes. How quickly can change to infrastructure be eliminated as the cause of a problem? Can all changes be mapped to authorized work orders? Studies show that up to eighty percent of all systemic outages are the result of authorized employees making changes to infrastructure. We also know that nearly eighty percent of the time it takes to solve a problem is consumed by the process of pinpointing the location and nature of the problem.

We also examine the organization's rollback capabilities for use in problem management. Here we look at how quickly staff can return the infrastructure to a known, good state as it existed before a problem occurred. This allows quicker restoration of services when there is an outage and also allows senior staff to examine the forensic evidence from the failure away from the heat of battle.

Release Management Processes

In release management we are looking for repeatable builds capability for any piece of critical infrastructure from scratch. This process is related to configuration management where the configuration is stored and maintained. The ability to quickly recreate the last known good configuration is critical to your organization's ability to respond to incidents such as disaster, security breach, major vendor failure or the outbreak of a virus or worm. It is also crucial when deploying software upgrades and patches that fail or display bugs. The assessment also looks at acceptance processes for new software and hardware to determine if they are ready for mass deployment into actual production capacity. This guards the integrity of both the configuration management and repeatable build processes by insuring that no new infrastructure or software is deployed that can't be rebuilt or controlled by configuration management and that hasn't been reviewed by stakeholders such as Security.

Security

The final area we evaluate is security. Security is not a task that is done but rather the end-result of many other processes and controls operating effectively. Here, we examine your organization's ability to understand the known good state of IT infrastructure, how it is configured and built, and how to know if someone makes a malicious change versus a change made by internal staff. If an undesired change is made, can you rollback to a known good state instantly and provide forensic evidence that proves there was indeed malicious activity? How well have all of the critical pieces of infrastructure been documented so that recovery from a disaster is possible? Can policy circumvention be tracked—is it possible to tell when change control policy has been violated? How easy is it for changes and new infrastructure to make it in to production without a prior security review.

Summary

In conclusion, the Integrity Management Capabilities Assessment was developed to assist organizations like yours to successfully implement IT process improvements and industry best practices in order to ensure the safety, stability and predictability of Information Services across your enterprise. The assessment is brought to you by the Information Technology Process Institute (ITPI), a non-profit organization with the unique charter to educate IT and move effective information Technology Service Management into the realm of responsible corporate governance. The ITPI serves to align actual practices with Best Practices and creates tools, prescriptive adoption methods and control metrics to facilitate management by fact.

The Integrity Management Capabilities Assessment was developed by the ITPI in partnership with IP Services and Tripwire. IP Services is a global, technical consulting organization whose focus is to help customers implement Managed Enterprise Network Operations Center (NOC) Services and E-Business solutions. Tripwire is the worldwide leader in integrity assurance solutions, delivering software for IT security and operations staffs so they can immediately detect and pinpoint undesired change to their servers and network devices. In this way, Tripwire enables rapid recovery, ensures the stability of information services, and increases systems availability as well as IT staff productivity.

Information Technology Process Institute
2896 Crescent Avenue, Suite 104
Eugene, Oregon 97408
Telephone: (541) 485-4051, opt. 5
Email: IMCA@itpi.org www.itpi.org

To schedule an Integrity Management Capabilities Assessment or to request additional information, please call (541) 485-4051.

V.E.E.S.C.

The value, effectiveness, efficiency, and security of IT controls: An empirical analysis

Information technology managers are confronted with a myriad of best-practice frameworks for information technology service management. These frameworks include the Information Technology Infrastructure Library (ITIL) and the Control Objectives for Information and related Technology (COBIT). Advocates of these frameworks promote the value of these guidelines in achieving cost reductions and improving business processes. The problem is that implementing these frameworks involves substantial upfront costs. Many practitioners view them as simply another level of bureaucracy. The purpose of this paper is to determine empirically whether IT controls affect the value, effectiveness, efficiency, and security of information-technology operations. We hypothesize that implementation of IT controls improves IT efficiency, IT effectiveness, IT security, and indirectly, business value. Based on prior research and extensive pilot testing with high-performing organizations, we are developing a survey to test our hypotheses. We will then distribute the survey to a sample of Fortune 1000 companies, government departments, and universities.

The benefits to you and your organization of completing the V.E.E.S.C. survey of practice include:

• Seeing where they rank nationally and by industry in terms of IT operational excellence.

  • Respondents will have free access to our overall IT ops excellence score calculations to continually review and rate their IT operations.

• Respondents will see evidence of the relationship between best practice IT controls and improved IT performance and return on IT investment.

  • The results will show the inter-relationships between the five BS 15000 process areas and their relative importance in determining IT performance.

  • Respondents will see how to improve their IT operational excellence ranking.

The V.E.E.S.C. benchmarking survey is a valuable addition to the information in Visible Ops: Starting ITIL in 4 Practical Steps. Do you want to participate or find out more information? Please contact us at veesc@itpi.org or visit http://www.itpi.org/home/veesc for more information.